Open Sourcing Space

With one exception, Silicon Valley lacks monumental landmarks that signify its importance as a world capital of technology innovation. That exception is Hangar One at Moffett Field in Mountain View, Calif., which is the home of NASA Ames Research Center.

Hangar One stands out like a Great Pyramid visible from Highway 101. Built to house airships called dirigibles or zeppelins, Hangar One opened in 1933. The floor inside this freestanding structure covers eight acres, and its enormous clamshell doors were designed for the passage of these airships. The hangar reputedly creates its own climate inside, bringing rain unexpectedly to parties that were organized there, back before it was closed.

Today, the future of this historic structure depends on NASA and various groups debating whether to restore it or tear it down. (Its walls are covered with siding that contains asbestos and PCBs.) Those who would preserve it recognize its power as a cultural symbol. While the days of airships are mostly gone — Airship Ventures now runs zeppelin tours from Moffett Field — Hangar One remains an inspiration.

Inspiration was a by-product of the space race in the United States. Many, like me, thought of themselves as part of the space program, following the Mercury and Apollo missions, even if our role was limited to watching TV. The goal of a moon landing inspired young men and women to become scientists and engineers. They entered NASA with great enthusiasm to become part of something as big as they could imagine.

Many had satisfactory careers inside NASA, while others grew frustrated as NASA became a slow-moving bureaucracy. Increasingly, NASA made it harder (and more costly) to do anything. So, like the age of dirigibles, the U.S. space program that I grew up with is gone, and like Hangar One, its future is uncertain. Yet our fascination with space is not.

One cause for hope is that the future of space exploration doesn't depend solely on NASA. Bruce Pittman, who works in the Space Portal group at NASA Ames, calls this future "Space 2.0." If Space 1.0 was a "one-size-fits-all" approach with

> The U.S. space program that I grew up with is gone. Yet our fascination with space is not.

NASA controlling everything, Pittman says, then Space 2.0 depends upon "massive participation," harnessing enthusiasm and expertise in places around the globe.

Space 2.0 represents the open sourcing of space exploration, a new model that could lead to faster, cheaper ways to develop space technologies.

It's also a call for makers to participate in research and development. Just as we're seeing amateurs play a role in astronomy and other fields, amateurs will be undertaking projects in support of a next-generation space program. For example, Lynn Harper of NASA Ames points out that the commercialization of space will mean a huge increase in suborbital flights, and a growing field of research in microgravity. She says this research requires "not just hundreds of experiments to send into space, but hundreds of thousands."

In this "DIY Space" issue of MAKE, you'll meet all kinds of makers, some inside NASA but many more outside the agency. We look at how to build your own homebrew satellites that take payloads into near-space and even into orbit. We show you how to build fast, cheap gadgets to analyze galactic spectra or eavesdrop on the space station. We also look at a variety of space-related projects seeking the participation of makers like you, from smart-phone satellites to lunar mining robots.

For his report "Rocket Men," Charles Platt interviewed the makers of a new private space industry. He also visited the Mojave Air and Space Port, where individuals and small companies set up to do space research. Spaceport manager Stuart Witt says, "I offer people the freedom to experiment." That's all you really need. The future, if you're so inspired, is up to you.

Dale Dougherty is the founder and general manager of Maker Media.

Volume 24
Make:
technology on your time

SPACE

ON THE COVER: Lego in (simulated) orbit.
With a Lego Mindstorms NXT system, students
and engineers at NASA Ames built a full-function
prototype satellite. Photograph by Garry McLeod,
background by Corbis. More NASA Ames photos at
makezine.com/24/space.

Columns

DIY UP HIGH:
Launch a balloon
with a scientific
payload to
the stratosphere.

Vol. 24, Oct. 2010. MAKE (ISSN 1556-2336) is published quarterly by O'Reilly Media,
Inc. in the months of January, April, July, and October. O'Reilly Media is located at
1005 Gravenstein Hwy. North, Sebastopol, CA 95472, (707) 827-7000. SUBSCRIP-
TIONS: Send all subscription requests to MAKE, P.O. Box 17046, North Hollywood,
CA 91615-9588 or subscribe online at makezine.com/offer or via phone at (866)
289-8847 (U.S. and Canada); all other countries call (818) 487-2037. Subscrip-
tions are available for $34.95 for 1 year (4 quarterly issues) in the United States;
in Canada: $39.95 USD; all other countries: $49.95 USD. Periodicals Postage Paid at
Sebastopol, CA, and at additional mailing offices. POSTMASTER: Send address
changes to MAKE, P.O. Box 17046, North Hollywood, CA 91615-9588. Canada
Post Publications Mail Agreement Number 41129568. CANADA POSTMASTER:
Send address changes to: O'Reilly Media, PO Box 456, Niagara Falls, ON L2E 6V2

Make: Projects

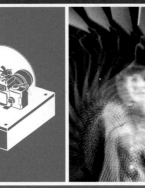

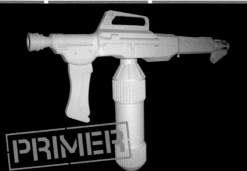

Can You Solve This Puzzle?

FLASHLIGHT → **?** → **LED**

What is the missing component?

Make:

Volume 24

technology on your time

Maker

PROTOTRIKE:
Inventor Saul Griffith takes his son, Huxley, for a ride in his tilt-steer cargo tricycle.

34

You can make it!

 Family fun project

Make:®
technology on your time

FOUNDER & GM, MAKER MEDIA
Dale Dougherty
dale@oreilly.com

EDITORIAL

EDITOR-IN-CHIEF
Mark Frauenfelder
markf@oreilly.com

MANAGING EDITOR
Keith Hammond
khammond@oreilly.com

SENIOR EDITOR
Goli Mohammadi
goli@oreilly.com

PROJECTS EDITOR
Paul Spinrad
pspinrad@makezine.com

STAFF EDITOR
Arwen O'Reilly Griffith

EDITOR AT LARGE
David Pescovitz

CREATIVE DIRECTOR
Daniel Carter
dcarter@oreilly.com

DESIGNER
Katie Wilson

PRODUCTION DESIGNER
Gerry Arrington

PHOTO EDITOR
Sam Murphy
smurphy@oreilly.com

COPY EDITORS
Gretchen Bay
Bruce Stewart

EDITORIAL ASSISTANT
Laura Cochrane

PUBLISHING

MAKER MEDIA DIVISION
PUBLISHER
Fran Reilly
fran@oreilly.com

SALES DEVELOPMENT MANAGER
Katie Dougherty Kunde
katie@oreilly.com

AD SALES MANAGER, EAST COAST
Selina Yee
212-964-8300
selina@oreilly.com

SALES ASSOCIATE
PROJECT MANAGER
Sheena Stevens
sheena@oreilly.com

CIRCULATION MANAGER
Sue Sidler

LOS ANGELES &
SOUTHWEST SALES
Jeff Griffith
Joe Hustek
626-229-9955

SAN FRANCISCO &
PACIFIC NORTHWEST SALES
Nick Freedman
707-775-3376

DETROIT & MIDWEST SALES
James McNulty
Mike Peters
248-649-3835

SINGLE COPY CONSULTANT
George Clark

ONLINE

DIRECTOR OF DIGITAL MEDIA
Shawn Connally
shawn@oreilly.com

DIRECTOR OF TECHNOLOGY
Stefan Antonowicz
stefan@oreilly.com

WEB DEVELOPER
Madelin Woods

EDITOR-IN-CHIEF
Gareth Branwyn
gareth@makezine.com

SENIOR EDITOR
Phillip Torrone
pt@makezine.com

ASSOCIATE EDITOR
Becky Stern

COMMUNITY MANAGER
Matt Mets

E-COMMERCE

ASSOCIATE PUBLISHER & GM,
E-COMMERCE
Dan Woods
dan@oreilly.com

DIRECTOR, RETAIL MARKETING
& OPERATIONS
Heather Harmon Cochran

OPERATIONS MANAGER
Rob Bullington

MAKER SHED PRODUCT
DEVELOPMENT
Marc de Vinck

EVENTS

DIRECTOR, MAKER FAIRE,
& EVENT INQUIRIES
Sherry Huss
707-827-7074
sherry@oreilly.com

MAKER FAIRE SALES & MARKETING
COORDINATOR
Brigitte Kunde
brigitte@oreilly.com

MAKE TECHNICAL ADVISORY BOARD
Kipp Bradford, Evil Mad Scientist Laboratories, Limor Fried, Joe Grand, Saul Griffith, William Gurstelle, Bunnie Huang, Tom Igoe, Mister Jalopy, Steve Lodefink, Erica Sadun, Marc de Vinck

CONTRIBUTING EDITORS
William Gurstelle, Mister Jalopy, Brian Jepson, Charles Platt

CONTRIBUTING WRITERS
Tim Anderson, John Baichtal, Tom Baycura, Chris Boshuizen, Jared Bouck, Nicole Catrett, David Cline, Abe Connally, Larry Cotton, Michael Covington, Sharon Covington, Cory Doctorow, Peter Edwards, Diana Eng, Alan Federman, Simon Quellen Field, Adam Flaherty, Thomas Fox, Saul Griffith, Bob Harris, Joshua Hart, Rachel Hobson, Steve Hoefer, Jon Kalish, Laura Kiniry, Bob Knetzger, Andrew Lewis, Forrest M. Mims III, James Newell, Meara O'Reilly, Tom Parker, John Pavlus, Eric Ponvelle, Kevin Quiggle, Matthew F. Reyes, Kristin Roach, Adam Sadowsky, Adam Savage, Terrie Schweitzer, L. Abraham Smith, Bruce Stewart, Cy Tymony, Ariel Waldman, Megan Mansell Williams, Matthew Wirtz, Edwin Wise, Lee David Zlotoff

CONTRIBUTING ARTISTS
Nick Dragotta, Tim Lillis, Garry McLeod, Rob Nance, Damien Scogin, Jen Siska

ONLINE CONTRIBUTORS
John Baichtal, Chris Connors, Collin Cunningham, Adam Flaherty, Kip Kedersha, Matt Mets, John Edgar Park, Sean Michael Ragan, Marc de Vinck

INTERNS
Eric Chu (engr.), Brian Melani (engr.), Tyler Moskowite (engr.), Ed Troxell (photo), Nick Raymond (engr.)

PUBLISHED BY
O'REILLY MEDIA, INC.
Tim O'Reilly, CEO
Laura Baldwin, COO

Copyright © 2010
O'Reilly Media, Inc.
All rights reserved.
Reproduction without
permission is prohibited.
Printed in the USA by
Schumann Printers, Inc.

CUSTOMER SERVICE
cs@readerservices.
makezine.com

Manage your account online,
including change of address:
makezine.com/account
**866-289-8847 toll-free
in U.S. and Canada
818-487-2037,
5 a.m.–5 p.m., PST**

Visit us online:
makezine.com

Comments may be sent to:
editor@makezine.com

Contributors

Dr. Chris Boshuizen (*Make Your Own Satellites*) is a contractor at NASA Ames Research Center at Moffett Field, Calif., where he serves as small spacecraft technical liaison, interfacing with NASA centers and commercial partners on the use of Ames-developed technologies. Chris is leading development of new, super cheap approaches to space exploration. "My goal," he says, "is to make space travel as easy as catching a bus." With a Ph.D. in physics from the University of Sydney, he's a former director of the Space Generation Advisory Council of students and young professionals, and was a co-founder and interim director of Singularity University. He has always wanted to be an astronaut.

Joshua Hart (*Cargo Bikes*) is a Northern California writer and activist working on transportation, climate, and wireless radiation issues. "My life has been enriched more by what I've given up than what I consume," he reports. He no longer owns a car, flies on planes, smokes, or uses cellphones or wi-fi. It might sound like deprivation, but, he says, "When it comes to food as well as transportation, one thing I've found is that slow is better! I am happier and healthier than I have ever been in my life." onthelevelblog.com

Diana Eng (*Yagi Antenna*), aka operator KC2UHB, is a fashion designer who works with math, science, and technology. She also writes about ham radio for MAKE and Make: Online. While busy starting her own fashion line (dianaeng.com) and developing her first product (a fortune cookie coin purse!), she also likes to "climb up rooftops in New York City to communicate through satellites using amateur radio." Her experiences as a designer have been diverse: sitting front row at New York Fashion Week, being a designer on TV's *Project Runway*, and co-founding the NYC Resistor hacker group. She lives in Brooklyn, N.Y., with her fiancé, Dave Clausen, aka operator W2VV.

Bob Knetzger (*Toy Inventor's Notebook*) is an independent designer and inventor with a background in industrial design. He's had fun inventing toys and games at Mattel, designing graphics for Intellivision games, creating educational software, and, with his Neotoy partner Rick Gurolnick, developing new product lines like Doctor Dreadful. Bob lives near Seattle with his wife, Deborah, and plays pedal steel guitar and banjo when he's not working on a new, top-secret toy!

Artist and musician **Peter Edwards** (*DIY Volume Control*) is a "self-taught electronics tinkerer and full-time maker." He lives in upstate New York with his girlfriend and two cats, and makes a living building musical instruments (casperelectronics.com) and occasionally teaching workshops. Lately he's selling a synth kit called the Drone Lab and watching Japanese horror movies. What's on his mind? "I've been crazy lately about high-power tricolor LEDs. You can create very dramatic effects that I never would have imagined I could get from a little LED." So now he's working on a new sound-to-light circuit that uses a microcontroller to output a responsive, colored light display.

Nick Raymond's (MAKE engineering intern) interests in engineering and the outdoors collide at Doran Beach in Sonoma County, Calif., where he works part time and surfs. "I became fascinated with the concept of harnessing the power of the ocean to convert to electrical power," he says. Since that fascination took hold, he has built a wave energy converter, and is currently building a DIY wave tank. The president of the Santa Rosa Junior College Engineering Club, Nick likes the color blue, sushi, and his benchtop mini mill. He hopes to work on projects "that will one day provide renewable and clean energy for the populations of the world, or something like that."

Robomowers, web telegraphy, idea sharing, and MAKE saves the day.

✉ Your "Lawnbot" cover article by J.D. Warren (Volume 22) got me so fired up that I just had to build one of my own! I took some different paths, using an off-the-shelf motor controller and a battery-powered mower that I can switch on and off remotely, through an Arduino-compatible board.

Not only does my "Robomower" keep my lawn looking the best it ever has, but it stops traffic! The best part: one out of every three or four passersby asks, "Do you ever read MAKE magazine?"

This has been one of my most fun and rewarding summer projects ever (see my build log and video at makezine.com/go/robomower). From my comfy lawn chair, my glass is raised high to you and J.D.!
—*Michael Zenner, Portland, Ore.*

✉ It's great that Saul Griffith's mailbox is overflowing with ideas from generous makers (Volume 23, "Smiley Face Technologies") but wouldn't it be even better to share these ideas online with the rest of us? I'll even make the website! That would be the ultimate Smiley Face Technology.
—*Joey Mornin, Berkman Center for Internet & Society at Harvard, Cambridge, Mass.*

✉ I thoroughly enjoyed Volume 23, except for one little point in "Remaking History." Telegraph sounders are intentionally constructed to give a different sound on closing and on opening; the relative time between "tik" and "tok" signifies the complete dot or dash. This one's a nice relay but not a true sounder.

That said, Bill Gurstelle's a good technical writer and MAKE is a good book. I not only derive the pleasure of learning from it (I've been back to Volume 23's "Primer: Programmable Logic Controllers" about five times) but as a journalism professor who teaches freelance writing, I use MAKE as an example.

By the way, a great number of us connect on the web with our antique telegraph equipment — I often have MorseKOB software (morsekob.org) running in the background, with a vintage U.S. Army Signal Corps "bug" key and an old Menominee sounder hooked up to my computer.
—*Charles A. Hays, Ph.D. aka WB7PJR/VE7 aka "CH" on the wire Thompson Rivers University, Kamloops, B.C.*

✉ Thank you for instantly confirming the usefulness of my new subscription. The "iPhone Screen Repair" how-to (Volume 22) was spot-on and saved me hours of waiting in line for an iPhone 4!
—*Adam Velez, San Juan Capistrano, Calif.*

✉ I'm a charter subscriber, and unfortunately, a flooded basement recently destroyed my Volumes 01 to 05. When I visited the MAKE Digital Edition, I was so pleased to see they were now available as downloadable PDFs — it came close to a teary-eyed moment. Now I can read them (again), and being able to print out those project pages is wonderful!
—*Clark Kielian, North Brunswick, N.J.*

MAKE AMENDS

In Volume 23's "Electronics: Fun and Fundamentals" column, the descriptions for the 4078 chip's pins 8 and 13 were mistakenly reversed in Figure F(d) on page 143. Pin 13 is the NOR output and pin 8 is unused.

In Volume 23's "Squelette, the Bare Bones Amplifier," the wrong perf board was specified; the right one is RadioShack #276-147, which has solder pads on one side and measures 4½"×6⅝". Also, the total cost was estimated at under $50, but this figure assumed that you already have common components (resistors, capacitors, etc.) on hand. The cost is higher if you buy everything new and in small quantities.

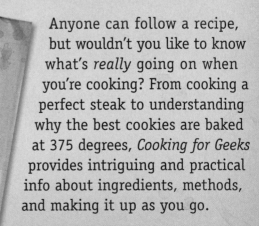

Model It, or Make It Modelable

This summer I had two great interns in my lab. As usual, they taught me more than I taught them. One in particular helped me refine my thoughts on the theme that pervades my every day.

When Geoff first arrived in my office, I described his summer project to him. I had written 2,000 lines of code in MATLAB to model the Ackerman steering geometry of a tilting (leaning) vehicle. He was to take that code, check it, improve it, and finish it, and we'd build the vehicle as a tilting, steering, cargo-carrying tricycle. He said, "It'll take three days." I countered, "I'll bet it takes six weeks."

Geoff dove into the code. He only looked up from his computer for two reasons: to hear instructions for using the vintage hand-pulled espresso machine, and to go stare at the physical prototype of the tilting trike to orient himself to the problem. He missed his own three-day target, but crushed my six-week estimate when he proudly showed me the first working computer model in just two weeks.

And that's the theme pervading my whole life right now: computational modeling.

Why? Here's how I see it: Galileo Galilei arguably did more to usher in the scientific revolution than any other. The quote "Measure what is measurable, and make measurable what is not so" is attributed to him. In my mind, 19th- and 20th-century science did exactly that, and the scientific method — the cornerstone of thoughtful progress in knowledge — is heavily dependent on good measurement.

We measured everything we could about the scale of the universe. We probed the atomic then subatomic structure of the elements with incredibly elegant, single-parameter experiments that isolated things like the mass of an electron.

This type of science has been so successful that now it seems the true frontiers of science exist less in the study of easily reducible, measurable things, and more in the study of complex, multiparameter systems — like biology, climate, metabolism, and ecology. In these systems, understanding is built with models that can be tested for their validity and correspondence to the messy, complex real world.

Like many physical systems, there's no perfect answer to a tilting tricycle, only "optimal." You can optimize the parameters, but because of the limits of physically realizable machines, it can only ever be "almost perfect." Success is being closer to almost perfect than other people's models.

But here's the beautiful thing about modeling. Computational models are digital, and that makes them inherently shareable, independently verifiable, and easy to collaborate on and improve.

Whereas my inclination was to immediately start to build something physical, Geoff's approach — the approach of a new generation of engineers and scientists — was to begin with a model. Start with bits. Make them perfect, beautiful, defendable, sharable bits, then render them physical once you've reached an optimum. Sure, someone might figure out a better optimum one day, but because they can start with working, executable code, they'll get to it faster.

There's an even more important reason to encourage this culture of shared models. The more people who have experience simulating the world with success, and making things from those models, the more people will trust in the models of our physical world that will guide how we shape our future.

I read once about the science of perception and the humble practice of catching a falling ball. A ball moving 60mph travels almost 90 feet in a second. The only reason we can catch it is because we have a mental model of where the ball will be when our hand intercepts it. Throughout the course of our lives, we've built a mental computational model, which we've refined thousands of times, that helps us predict the future position of a ball so we might catch it with our relatively slow reflexes.

We have enormous faith in the ability of a professional baseball player to model the future of a ball, under complex windy, rainy, and noisy conditions, and to catch it. It would be nice to build similar public faith in the ability of our professional scientists to model the future — the future of the oceans if we continue to pollute them with toxins, of the atmosphere if we continue to emit carbon dioxide, and of other problems that require humanity to have a faster response time than its cultural reflexes.

Saul Griffith is a new father and entrepreneur. otherlab.com

MAKER'S CALENDAR

COMPILED BY WILLIAM GURSTELLE

Our favorite events from around the world.

Space Shuttle *Discovery* Launch

Nov. 1, Cape Canaveral, Fla.

It's almost the last chance to experience the awesome power of the space shuttle's solid rocket boosters. The penultimate launch of the shuttle *Discovery* delivers critical parts and supplies to the International Space Station. nasa.gov/missions/highlights/schedule.html

NOVEMBER

» Nikon Photo Contest

Entries close Nov. 30, worldwide

This year's international photography contest includes both an energy-themed and a free-subject category. nikon-npci.com

» Wonderfest

Nov. 6–7, Berkeley, Calif.

Each fall, Wonderfest challenges science and technology enthusiasts by taking on tough or controversial topics in astronomy, biology, psychology, and physics. Activities include talks, discussions, science competitions, and a science expo. wonderfest.org

» Wings Over Homestead

Nov. 6–7, Homestead, Fla.

Restarted last year after a 17-year absence, the 2009 Homestead air show was packed with 200,000 people, and headlined by the U.S. Navy's Blue Angels. This year is expected to be even bigger. wingsoverhomestead.com

» The Woodworking Shows

Nov.–Mar., in 20+ U.S. cities

This traveling trade show tours the nation, bringing the latest in tools and techniques to winter-weary woodworkers. DIY seminars run the gamut from basic woodworking to cabinetmaking skills. thewoodworkingshows.com

DECEMBER

» Asia Game Show

Dec. 24–27, Hong Kong

Close to 400,000 visitors from around the world head to the Hong Kong Convention Center to discover the latest in online gaming and digital entertainment. This Christmastime event is a combination of entertainment, gaming technology, educational features, seminars, and hands-on gaming. asiagameshow.com

JANUARY

» North American International Motorcycle Supershow

Jan. 7–9, Toronto

Nearly 25 years old, the massive (425,000 sq. ft.) Supershow is North America's largest consumer motorcycle show. The exhibits span just about everything the motorcycle industry has to offer, from the latest bikes and accessories to choppers, scale models, DIY machines, and Bonneville racers. supershowevents.com

« Art Shanty Projects

Jan.–Feb., Minneapolis

It's part sculpture park, part artist residency, part social experiment. This temporary "shanty town" on a frozen lake was inspired by the traditional ice fishing houses that dot Minnesota in winter. A bit like a very small, very cold Burning Man. artshantyprojects.org

✱ IMPORTANT: Times, dates, locations, and events are subject to change. Verify all information before making plans to attend.

MORE MAKER EVENTS: Visit *makezine.com/events* to find events near you, like classes, exhibitions, fairs, and more. Log in to add your own events, or email them to *events@makezine.com*. Attended one of these events? Talk about it at *forums.makezine.com*.

Memento Mori

'm often puzzled by how *satisfying* older technology is. What a treat it is to muscle around an ancient teletype, feeding it new-old paper-tape or rolls of industrial paper with the weight of a bygone era. What pleasure I take from the length of piano roll I've hung like a banner from a high place in every office I've had since 2000.

How much satisfaction I derive from the racing works of the 1965 mechanical watch I received as a Father's Day present this year, audible in rare moments of ambient silence or when my hand strays near my ear, going tick-tick-tick-tick like the pattering heart of a pet mouse held loosely in my hand.

The standard explanation for the attractiveness of this old stuff is simply that They Made It Better In The Old Days. But this isn't necessarily or even usually true. Some of my favorite old technologies are as poorly made as today's throwaway products from China's Pearl River Delta sweatshops.

Take that piano roll, for example: a flimsy entertainment, hardly made to be appreciated as an artifact in itself. And those rattling machine-gun teletypes and caterpillar-feed printers — they have all the elegance of a plastic cap gun that falls apart after the first roll of caps has run through it.

Today, I have a different answer. Sitting beside me as I type this is a 512GB Kingston solid-state drive, its case lights strobing like the world's tiniest rave. Every time I look at this thing, I giggle. I've been giggling all afternoon.

I got my first personal computer in 1979, an Apple II+, and it came with 48K of main memory. I remember the day we upgraded the RAM to 64K, my father slotting in the huge board reverently, knowing that it represented $495 worth of our family's tight technology budget (about $1,500 in today's money). What I really remember is the screaming performance boost we got from that board.

The first time RAM made me laugh was in the mid-1990s. My mentor and friend, Miqe, and I were doing prepress jobs on brand-new Macintosh Quadras. It would often take a Quadra three or four days to complete a job. Of course, every machine already had as much RAM as it could handle (136MB).

Miqe and I got to talking about the performance

> The standard explanation for the attractiveness of older technology is simply that They Made It Better In The Old Days. But this isn't necessarily or even usually true.

improvements we'd be able to get with an unthinkable 500MB of RAM. Then we thought about 1GB of RAM and all we could do with it. Finally, we strained our imaginations to their outer limits and tried to imagine computing at 1TB of RAM.

And we started to laugh. This substance that cost more than its weight in gold — that solved all our problems — sometime in our lifetimes would be so cheap and abundant that we would have literally *unimaginable* amounts of it.

And that's why I've been giggling at this half-terabyte RAM (OK, RAM-like) drive that I just spent $1,500 on — the same sum Dad parted with for a 64K upgrade card 30 years ago.

Which brings me back to these beautiful old objects I have around my office. I don't have these here because they're inherently well-made. I have them because they're the best joke we have.

They're the continuous, ever-delightful reminder that we inhabit a future that rushes past us so loudly we can barely hear the ticking of our watches as they are subsumed into our phones, which are subsumed into our PCs, which are presently doing their damnedest to burrow under our skin.

The poets of yore kept human skulls on their desks as *memento mori* — reminders of mortality and humanity's fragility. I keep these old fossil machines around for the opposite reason: to remind me, again and again, of the vertiginous hilarity of our age of wonders.

Cory Doctorow's latest novel is *Makers* (Tor Books U.S., HarperVoyager U.K.). He lives in London and co-edits the website Boing Boing.

BUSINESS REPLY MAIL

FIRST-CLASS MAIL   PERMIT NO 865   NORTH HOLLYWOOD CA

POSTAGE WILL BE PAID BY ADDRESSEE

Make:
technology on your time

PO BOX 17046
NORTH HOLLYWOOD CA 91615-9588

Photography by Jason Dietz (jason.dietz@yahoo.com)

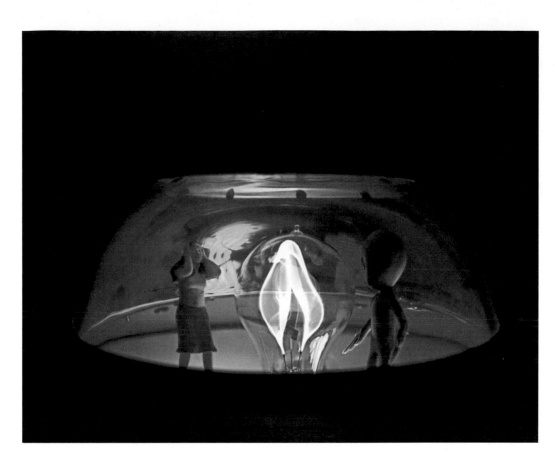

Warm Glow of Abduction

Inspired by the flying saucers, rocket ships, and robots of 1950s sci fi comic covers, **Jason Dietz** set out to create a little of that magic for his home. He decided to make lamps that depict a classic flying saucer shooting down a giant plasma ray and pulling up an unsuspecting victim into the ship. To get the desired effect, he knew he had to go big.

Dietz' UFO Lamps stand over 6 feet tall from base to saucer. The 2-foot-diameter flying saucer that crowns each lamp is a sturdy sandwich of parabolic aluminum heat dishes, Edison flame bulbs, and an acrylic disk. The saucer sits atop a giant hand-blown recycled-glass vase that holds 10 gallons of water.

CFL, LED, and halogen lights, in combination with a 110-volt air pump, nail the illusion, as the abduction victim, a lone cow, hovers and twirls helplessly above the grassy pasture from which it was plucked.

With its size, varied lighting, and constant motion, the lamp is beautiful and bizarre at once, not a sight easily overlooked. Dietz keeps one in his living room. "The soft glow of an alien abduction in progress in the corner of the room is quite the sight indeed," he says. "Staring at it for a while lets your imagination run wild — it puts me into that retro sci-fi world."

Like many makers, Dietz gains inspiration as much from seeing his visions come to life as from seeing others enjoying his creations. At Maker Faire Bay Area 2010, he displayed six of his UFO Lamps in a half circle at the back of Fiesta Hall, a dark environment that featured only projects that glow. Fairgoers were drawn in by the UFO beams, and thousands came closer for a good look.

"It looked like a small-scale alien invasion in the back of the hall," Dietz remembers. Apparently he wasn't the only one excited to see this fantasy made reality, as the lamps were in high demand.

His latest project is wall lights that integrate planets passing in front of each other in the manner of an eclipse. He hopes to display them at next year's Faire.

"We all have the power to create anything we want to see," says Dietz, "it just depends on how much you really want to see it happen." —*Goli Mohammadi*

➕ **Interview on Make: Online:** makezine.com/go/dietz

Crop Windows

Britta Riley grew up in Texas, so that might explain why she's growing bok choy in her Brooklyn loft. The 33-year-old maker believes that urban agriculture — specifically hydroponic "window farms" — can make a real contribution to environmental activism.

In 2010 Riley raised $27,000 for her Windowfarms project via the micro-donation website Kickstarter, and put together a system to grow plants in vertical columns hung in front of a window. She uses recycled spring water bottles, an aquarium air pump, air valve needles normally used to pump up basketballs, and hardware meant for hanging art.

"We're showing that you can actually get really far using things already available to us as consumers," says Riley.

Each column has four upside-down 1.5-liter plastic water bottles connected to each other; plants grow out of 4" holes cut in the sides with an X-Acto knife. The air pump circulates liquid nutrients that trickle down from the top of the column and wash over the plant roots.

More than 16,000 people have registered at Windowfarms' open source community website, and today there are window farmers around the globe, including Italy, Israel, Hong Kong, and Finland.

, The website has been crucial for exchanging ideas about improving the technology. It's a process Riley calls "R&DIY," or Research and Develop It Yourself. She cites as an example of R&DIY the window farmer who figured out a way to cope with the gurgling sound these systems make.

"He read up on gun silencers and then he just drilled a few holes into an empty vitamin bottle and stuck it over the end, and all of a sudden it completely silenced the system," reports Riley.

Windowfarms have been used to grow strawberries, cherry tomatoes, peppers, lettuce, and herbs. (You can't grow carrots, garlic, or other root vegetables with the system.) And for those who don't relish the opportunity to DIY entirely, a variety of Windowfarm kits are sold.

—*Jon Kalish*

≫ **Hydroponic Windowfarms:** windowfarms.org

Photograph by Ted Ullrich

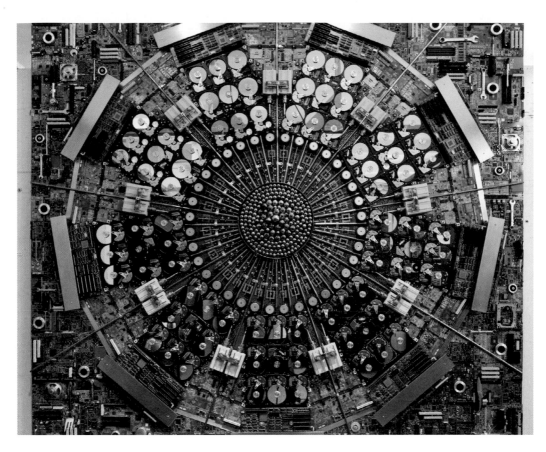

Photograph by Brad Plummer

Watching the Detectors

Deep underground beneath Switzerland and France, scientists at the Large Hadron Collider (LHC) at CERN are searching for particles that were present just moments after the Big Bang, when the universe formed. The massive machines built to conduct these experiments are astounding on their own — sheer marvels of precision and engineering. For one Bay Area artist, they're also his muse.

John Zaklikowski, 55, re-creates particle accelerator detectors like the ones at the LHC and at Fermilab in Illinois. What makes his large-scale assemblages so unique is what they're made of: computer motherboards, hard drives, video and sound cards, cellphone bits, vacuum tubes.

In his work *Large Hadron Collider*, based on the ATLAS and Compact Muon Solenoid detectors, the central feature consists of razor blades and a race-car air filter. "You've got to be very careful around this one," he warns. "When I was first making these things, I bled every day."

In *Fermilab*, Zak, as he prefers to be called, depicts a real collider detector with hard drives and motherboards, but also old telephone bells, cheese graters, and an ancient Chinese compass called a *luopan*.

Born and raised in Buffalo, N.Y., Zak studied philosophy and literature, not science. Sporting a full head of white hair, black jeans, and a pale blue Polo button-up splattered with paint, he says, "I did grapple with the idea of studying physics but didn't think it would really suit me. Obviously, my science interest is coming out in a big way in recent years with these pieces."

His work recently made the cover of the particle physics journal *Symmetry*, but Zak doesn't restrict his subject matter to physics experiments. There's also a baby grand piano in the middle of his San Francisco art studio, covered in parts. A newer work, *Enamored*, pays homage to his son, who, at 13, abandoned his collection of the Lego bricks. Luckily, his dad knew just what to do with the stash.

—*Megan Mansell Williams*

>> **Zak's Art in *Symmetry*:** makezine.com/go/zak

Free Juice Bar

The SolarPump Charging Station provides free power for charging cellphones, laptops, electric bikes, and other gadgets by converting solar energy to electricity right on the spot. Conceived by **Beth Ferguson**, an Austin-based designer, artist, and social entrepreneur, the SolarPump is meant to be both practical and thought-provoking.

Ferguson believes green charging stations are critical to helping society transition from gasoline to electric vehicles charged from renewable energy. "The SolarPump Electric Charging Station is designed to help people reimagine the future of transportation," she declares.

The first SolarPump used a 1950s Citgo gas pump to make sure the message really hit home. The power comes from solar panels on the roof, so there's no cost for charging. The face of the gas pump has an LED digital display that indicates solar panel voltage, total wattage of plugged-in devices, and battery bank energy level.

The current SolarPump station uses two deep-cycle 100-amp-hour, 12-volt batteries to create a 24-volt system, and an 1,100-watt inverter that supplies standard 110-volt AC power. Three bifacial solar panels charge the batteries through a charge controller. A Pentametrix system gathers data from the current shunts, then sends it digitally over RS-232 to the custom digital display module.

After building the first SolarPump prototype in 2009 for a design show at the University of Texas, Ferguson launched the nonprofit Sol Design Lab with a talented crew of designers, fabricators, and electricians. Their mission: to educate people about energy use and encourage a shift to electric vehicles, by exhibiting solar charging stations.

Four SolarPumps have been built so far, and they've been big hits at festivals and fairs like Coachella, South by Southwest, SolarFest, and Maker Faire. The latest charging stations incorporate outdoor furniture, lockers, and wireless internet.

—*Bruce Stewart*

>> **SolarPump Charging Stations:** soldesignlab.com
➕ **Interview on Make: Online:** makezine.com/go/solarpump

Photograph by Beth Ferguson

The Lawn Rider

Photograph by Matt Langley

Ted Wojcik is no stranger to bicycles — the 63-year-old former Navy aviation machinist is a legend in the cycling community for his custom hand-built frames — but he never anticipated being in the lawn care business. That's all changed with the Mow-Ped, a pedal-powered lawn mower that's since catapulted Wojcik to fame beyond the bicycling world.

The project first surfaced when Matt Langley, brother of former *Bicycling* tech editor Jim Langley, contacted Wojcik and commissioned him to build a lawn mower that was both environmentally and cyclist friendly. "He'd been using a mower marketed to be towed behind a wheelchair," says Wojcik, "and was pulling it along with an old mountain bike."

To create Mow-Ped, Wojcik enlisted help from his son, **Cody**, 24, a recent mechanical engineering graduate from Worcester Polytechnic Institute. Using 3D design software, they crafted a fully operational tractor before ever cutting a piece of metal.

Mow-Ped combines a tadpole-style recumbent bike, which has two front wheels and one in back, with go-kart-inspired steering. It accommodates riders ranging in height from 5' to 6'4". The manual reel mowers, situated directly beneath the seat, can be removed for the vehicle's recreational use.

Wojcik unveiled Mow-Ped at the 2010 North American Handmade Bicycle Show in Virginia to overwhelming interest, including environmentalists impressed with its ability to significantly reduce pollution (according to the EPA, 5% of all U.S. pollution comes from gasoline-powered lawn mowers).

Now with plans for large-scale production, the Wojciks are updating the original design: shortening the tractor for sharper turns, adding the capacity for fertilizing or seed drilling, and substituting wider mowers for more accurate cutting. They're also adding a standard fixed-gear rear hub so the Mow-Ped can work in reverse, with optional multi-speeds.

"It's not a mower for everyone," says Wojcik. "[But] if you're fit and have a yard that's somewhat flat and relatively obstacle-free, it works great."

—*Laura Kiniry*

≫ **Custom Bicycles:** tedwojcikcustombicycles.com

Aiming High, Real High

Daniel Parker's hobby isn't for everyone. He's devoted most of the last five years to designing and building his own high-altitude airplane.

Parker, 33, started flying in high school, joined the Experimental Aircraft Association at age 15, and built his first plane (an all-wood biplane) in college. Now he's hoping his current project might just beat the altitude record for small piston-engine aircraft.

Parker's background lends itself to such a complex hobby. He has a bachelor's in mechanical engineering and a master's in aeronautics and astronautics, both from Stanford. Taking time off from school to work at a composite airplane shop in Santa Monica, Calif., Parker was fortunate to find a friendly and skilled airplane builder/mentor, Dave Ronneberg.

Building a lightweight high-alt plane was attractive because it combined low mass, which Parker assumed meant less stuff to buy and build, with a definite yardstick for measuring success. Dubbed the Parker P1, the plane is made from aluminum and carbon fiber, and uses a Rotax 503 two-cylinder single-carb engine. The initial assembly began in Parker's one-bedroom apartment.

Parker's first priority is to conceptualize, build, and test-fly an airplane of his own design. But that's not to say he doesn't have his eye on a prize: the high-altitude record for the C1a0 weight class (under 661 pounds for plane, fuel, and pilot). The current record of just over 30,000 feet was set in 1989; Parker needs to beat it by 3% to get into the record books. At over 12,000 feet, supplemental oxygen is needed, so Parker's relying on the same system used by the U.S. Air Force.

The build has taken 6,000 hours and $40,000 so far, not counting workshop rent or specialized tools. But Parker says that it's really not all that difficult. "There were lots of frustrating moments, but I often say that there's no single process in building this plane that I couldn't teach anyone in an afternoon."

—*Bruce Stewart*

≫ **High Altitude:** parkerprojects.com/altitude.htm

Photograph by Dan Parker

Photograph by Andrew Carol

Lego Antikythera!

What would the ancient Greeks make of an iPod? According to **Andrew Carol**, they might have been more gadget-savvy than we give them credit for.

He's paid homage to their ingenuity by building a functioning Lego model of the famous Antikythera Mechanism — a 2,000-year-old handheld "mechanical computer" that people in Hipparchus' time used to make sophisticated astronomical predictions.

"Thousands of years ago, events like solar eclipses were terrifying to people," Carol says. "But they were smart — they recorded the patterns of when these eclipses occurred, and eventually some Greek guy realized he could build those patterns into a box of gears. After that, anyone could predict these important events just by turning a crank."

Carol builds software for Apple by day, but has explored the nexus "where computational mechanics and Lego meet" since 2006, when a cover story in an old *Scientific American* inspired him to build a working "difference engine" out of plastic bricks. (Charles Babbage's 19th-century device used cranks and gears to calculate mathematical functions.)

An editor at *Nature* saw Carol's model and asked if he could make a Lego Antikythera. Carol gladly accepted the challenge, but rebuilding antiquity's Palm Pilot was no small feat. "Whoever designed it had the luxury of cutting their own custom gears," he says, "whereas I just had to use what Lego makes."

Starting in late 2009, he designed a modular system comprising seven mechanical differentials and more than 100 gears to achieve the "exotic ratios" necessary for computing lunar movements. Two prototypes, 10 days of Christmas vacation, and $500 worth of Lego Technic pieces later, Carol had a working version of the machine.

He recently presented it to adoring geeks at the annual Science Foo Camp held at Google's headquarters in Mountain View, Calif. At roughly the size of a desktop printer, his replica "isn't as compact as the original," he concedes. "Then again, I had to use twice as many gears as they did." —*John Pavlus*

>> **Andrew Carol's Work:** acarol.woz.org
➕ **Antikythera Device:** makezine.com/go/antikythera

Becoming an Amateur Scientist

A n editorial in a leading science journal once proclaimed an end to amateur science: "Modern science can no longer be done by gifted amateurs with a magnifying glass, copper wires, and jars filled with alcohol." I grinned as I read these words. For then as now there's a 10x magnifier in my pocket, spools of copper wire on my workbench, and a nearby jar of methanol for cleaning the ultraviolet filters in my homemade solar ultraviolet and ozone spectroradiometers. Yes, modern science uses considerably more sophisticated methods and instruments than in the past. And so do we amateurs. When we cannot afford the newest scientific instrument, we wait to buy it on the surplus market or we build our own. Sometimes the capabilities of our homemade instruments rival or even exceed those of their professional counterparts.

So began an essay about amateur science I was asked to write for *Science* (April 1999, bit.ly/cTuHap), one of the world's leading science journals. Ironically, the quotation in the first sentence came from an editorial that had been previously published in *Science*.

In the 11 years since my essay appeared in *Science*, amateur scientists have continued doing what they've done for centuries. They've discovered significant dinosaur fossils, found new species of plants, and identified many new comets and asteroids. Their discoveries have been published in scientific journals and books. Thousands of websites detail an enormous variety of amateur science tips, projects, activities, and discoveries. Ralph Coppola has listed many of these sites in "Wanderings," his monthly column in *The Citizen Scientist* (sas.org/tcs).

Today's amateur scientists have access to sophisticated components, instruments, computers, and software that could not even be imagined back in 1962 when I built my first computer, a primitive analog device that could translate 20 words of Russian into English with the help of a memory composed of 20 trimmer resistors (bit.ly/atF5VL).

Components like multiwavelength LEDs and laser diodes can be used to make spectroradiometers and instruments that measure the transmission of light through the atmosphere. Images produced by digital video and still cameras can be analyzed with free software like ImageJ to study the natural world in ways that weren't even imagined a few decades ago. Amateur astronomers can mount affordable digital cameras on their telescopes, which then scan the heavens under computer control.

Cameras, microscopes, telescopes, and many other preassembled products can be modified or otherwise hacked to provide specialized scientific instruments. For example, digital camera sensors are highly sensitive to the near-infrared wavelengths beyond the limits of human vision from around 800nm–900nm. IR-blocking filters placed over camera sensors block the near-IR so that photographs depict images as they'd be seen by the human eye. Removing the near-IR filter provides a camera that can record the invisible wavelengths reflected so well by healthy foliage.

Many of the makers who publish their projects in the pages of MAKE, *Nuts and Volts*, and across the web have the technical skills and resources to devise scientific tools and instruments far more advanced than anything my generation of amateur scientists designed. They also have the ability to use these tools to begin their own scientific measurements, studies, and surveys. Thus, they have the potential to become the pioneers for the next generation of serious amateur scientists.

Previous installments of this column have covered approaches for entering the world of amateur science, and future columns will present more. For now I'll end this installment with a brief account of how I began doing serious amateur science so you can see how a relatively basic set of observations of the atmosphere has lasted more than 20 years and, with any luck, will continue for another 20 years.

Case Study: 20 Years of Monitoring the Ozone Layer

In May 1988, I read that the U.S. government planned to end a solar ultraviolet-B radiation monitoring program due to problems with the instruments. Within a few months I began daily UVB monitoring using a homemade radiometer. The radiometer used an

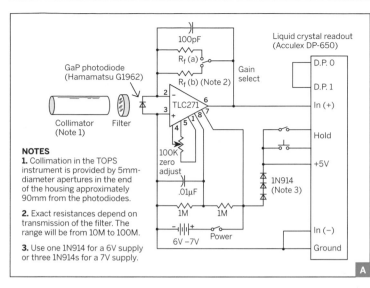

Photography and images by Forrest M. Mims III (A, B); Goddard Space Flight Center, NASA (C)

Fig. A: Circuit diagram for one of the two UVB sun photometers in the original homemade TOPS-1 from 1990. All the components are still available. Fig. B: The TOPS project earned a 1993 Rolex Award that provided funds for the development of a first-generation micro-processor-controlled TOPS (Microtops) by Scott Hagerup. Fig. C: This global ozone image was acquired while NASA's Nimbus-7 satellite was providing accurate data during 1991. On this day TOPS-1 measured 284.4 Dobson units (DU) of ozone, and the satellite measured 281.5 DU.

NOTES

1. Collimation in the TOPS instrument is provided by 5mm-diameter apertures in the end of the housing approximately 90mm from the photodiodes.

2. Exact resistances depend on transmission of the filter. The range will be from 10M to 100M.

3. Use one 1N914 for a 6V supply or three 1N914s for a 7V supply.

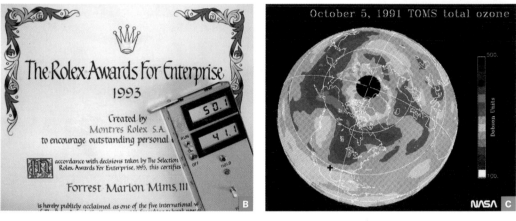

inexpensive op-amp integrated circuit to amplify the current produced by a UV-sensitive photodiode. An interference filter passed only the UVB wavelengths from about 300nm–310nm, while blocking the visible wavelengths.

I described how to make two versions of the UVB radiometer in "The Amateur Scientist" column in the August 1990 *Scientific American*. This article also described how the radiometer detected significant reductions in solar UVB when thick smoke from forest fires at Yellowstone National Park drifted over my place in South Texas in September 1988.

Ozone strongly absorbs UV, and the amount of ozone in a column through the entire atmosphere layer can be determined by comparing the amount of UV at two closely spaced UV wavelengths. This is possible because shorter wavelengths are absorbed more than longer wavelengths.

This meant that my simple UVB radiometer formed half of an ozone monitor. So I built two radiometers inside a case about half the size of a paperback book. One radiometer's photodiode was fitted with a filter that measured UVB at 300nm, and the second was fitted with a 305nm filter. I named the instrument "TOPS" for Total Ozone Portable Spectrometer. (Full details are at bit.ly/9JOth9.)

TOPS was calibrated against the ozone levels monitored by NASA's Nimbus-7 satellite. This provided an empirical algorithm that allowed TOPS to measure the ozone layer to within about 1% of the amount measured by the satellite. During 1990, ozone readings by TOPS and Nimbus-7 agreed closely. But in 1992, the two sets of data began to diverge so that TOPS was showing several percent more ozone than the satellite.

When I notified the ozone scientists at NASA's

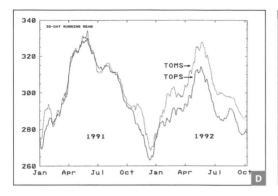

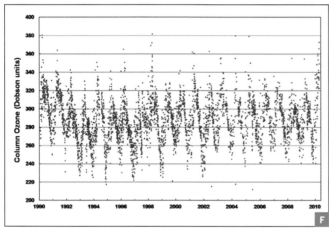

Fig. D: *Nature* plot compares ozone measurements by TOPS and Nimbus-7. In 1992 the calibration of the satellite's instrument began to drift. Fig. E: Scientist Brooke Walsh measures the ozone layer with the world-standard ozone instrument at Hawaii's Mauna Loa Observatory. Fig. F: The ozone layer over South Texas, measured by the author. Red points from 1990 to 1994 were measured by TOPS-1. Blue points from 1994 to 1997 were measured by Microtops and Supertops. Points from 1997 to 2010 were measured by Microtops II, manufactured by Solar Light.

Goddard Space Flight Center (GSFC) about the discrepancy, they politely reminded me that the satellite instrument was part of a major scientific program and not a homemade instrument. I responded that I had built a second TOPS and both showed a similar difference, but this didn't convince them.

During August of 1992, I visited Hawaii's Mauna Loa Observatory for the first time to calibrate my instruments at that pristine site 11,200 feet above the Pacific Ocean. The world-standard ozone instrument was also being calibrated there, and it indicated a difference in ozone measurements made by Nimbus-7 that were similar to what I had observed.

Eventually NASA announced that there was indeed a drift in the calibration of its satellite ozone instrument. A paper I wrote about this sparked my career as a serious amateur scientist when it was published in *Nature*, another leading science journal ("Satellite Ozone Monitoring Error," page 505, Feb. 11, 1993). Later GSFC invited me to give a seminar on my atmospheric measurements that they titled "Doing Earth Science on a Shoestring Budget." That talk led to two GSFC-sponsored trips to study the smoky atmosphere over Brazil during that country's annual burning season, and several trips to major forest fires in western U.S. states.

Going Further

The regular ozone measurements I began on Feb. 4, 1990, have continued to this day along with measurements made by various homemade instruments of the water vapor layer, haze, UVB, and other parameters. In future columns we'll explore how you can also make such measurements — and very possibly make discoveries of your own.

Forrest M. Mims III (forrestmims.org), an amateur scientist and Rolex Award winner, was named by *Discover* magazine as one of the "50 Best Brains in Science." He edits *The Citizen Scientist* (sas.org/tcs).

THE INVASION HAS BEGUN

uat.edu/robotics
877.UAT.GEEK

ROBOTS ARE TAKING OVER.

LEARN, EXPERIENCE AND INNOVATE WITH THE FOLLOWING DEGREES: Advancing Computer Science > Artificial Life Programming > Digital Media > Digital Video > Enterprise Software Development > Game Art and Animation > Game Design > Game Programming > Human-Computer Interaction > Network Engineering > Network Security > Open Source Technologies > Robotics and Embedded Systems > Serious Game and Simulation > Strategic Technology Development > Technology Forensics > Technology Product Design > Technology Studies > Virtual Modeling and Design > Web and Social Media Technologies

Bug Frog (UC)

A Rube Goldberg Music-Making Machine

Musical group OK Go released the video for their song "Here It Goes Again" in 2005. The iconic video shows the four band members dancing wonderfully on eight moving treadmills. Shot with a single locked-off camera, and now exceeding 50 million views on YouTube, that video redefined what a viral video could be.

In August 2009, Syyn Labs began discussions with the band to build them a machine they could "dance with" in a Rube Goldberg-style chain reaction for their next video. A few requirements: no "magic," and the machine should try to hit beats throughout, play part of the song, and be built to be photographed in one continuous shot.

The build was daunting, but ultimately it was a great success. (See the video at bit.ly/okgosyyn.) Here are some things we learned.

1. Do the small stuff first. We wanted the machine to build in excitement, to crescendo with lots of big, crazy interactions. Also, we've found that …

2. Bigger is better. Smaller components are more fidgety than larger objects. A marble and its trigger are simply far more affected by dirt, temperature changes, and vibration than a bowling ball, which doesn't much care at this scale. Therefore we also …

3. Put the less reliable stuff up front. This was important, so that we spent as little of the precious shooting time resetting the machine as possible. With 89 different types of interaction, and many times more than that if you count each physical interaction (each domino, chair, rat-trap flag, etc.), we wanted any failures to happen near the beginning. Nevertheless, it's important to …

4. Have lots of people involved. Ultimately, we had more than 55! They were all essential and worked long hours late into the night to get everything working beautifully. Of course, when you have that many people working on a machine so large, you must …

5. Assign teams. This machine was really big, so having a dedicated team for key components helped improve reliability and minimize danger (from falling pianos and steel drums). This specialization also allowed for flexibility when last-minute changes were needed. Why the changes? Well …

6. Aesthetics are important. Some interactions in the machine were too fast, or too crowded, or just didn't have the right "feel" on camera. For example, the piano was intended to come down slowly, but it looked so good crashing down that we decided to make that change. Unintended consequences occured due to vibration, but a workaround was discovered, and we continued, proving my last point …

7. Be flexible. Early on, we knew we wanted to end with paint cannons, and we suspected we'd start with dominoes (a classic!), but the rest was a realm of infinite possibilities. Many ideas were pursued and abandoned: some were hard to photograph, others simply didn't work. It's hard to give up on an idea after days or even weeks of work are invested, but, we found, it's often necessary for the greater good.

➕ More photos at: makezine.com/24/learned

Adam Sadowsky is president of Syyn Labs (syynlabs.com), a Los Angeles-based collective of interdisciplinarily talented scientists, engineers, designers, and technologists who love to explore the blurry distinction between art and technology.

Photography by Edwin Roses

Garden of DIY Delights

Make: Projects (**makeprojects.com**) **is our newest online service, a living library of how-to** tutorials being built by the entire maker community. Sure it's got great projects from our magazine and websites, but more importantly it's got great projects contributed by makers like you, plus wiki-based primers ranging from cooking and brewing to gear-cutting and servomotor basics.

Because all of Make: Projects is built on a custom wiki (thanks to our friends at iFixit), every article is a "living document" that can be edited by anybody. Have some expertise? Share it by writing a how-to. Find a mistake? You can fix it. Did you do a project and take better photographs? Add them. Have questions or feedback about a build? Add those too, and see the project improve.

We're excited to see how this service grows and changes over time. To make it truly great, we need your help. You can get started at makeprojects.com/Info/start. Here's a sampling — what will you contribute?

PROJECTS FROM MAKE

Step-by-step projects from MAKE magazine and Make: Online.

☼ Geared Candleholder

Kinetic sculptor Ben Cowden shows how to make a burly, elegant candleholder with two gear-driven arms that raise and lower the candle platforms. You'll learn to fabricate your own gears,  using sheet aluminum and a drill to cut the teeth. From MAKE Volume 21. makezine.com/go/geared

☼ Medicine Man Glider Build from scratch an old-school, balsa-wood-and-tissue-paper glider with an awesome 5-foot wingspan. Much larger photos than appeared in the magazine (Volume 17) make the details of this involved build easier to see and understand. makezine.com/go/glider

USER-CONTRIBUTED PROJECTS

Makers are rolling up their sleeves and sharing their how-tos and DIY wisdom with us — keep 'em coming!

☼ Install a Penny Countertop Replace a boring counter surface with a lustrous, durable, and cheap new top using two-part clear epoxy, marine-grade polyurethane, and a sack of pennies. makezine.com/go/penny

☼ Mylar Light Box Ryan Jenkins of the Exploratorium shows how rolls of mylar film in a box, covered with tissue paper, can produce all sorts of beautiful patterns when held over a light source. And check out his fantastic photos. makezine.com/go/mylar

TECHNIQUES AND PRIMERS

Over time, we expect the collected expertise of the maker community to become a significant draw.

☼ Technique: Label-Etch a Glass Bottle Sean Ragan shares a simple trick he discovered for etching designs onto glass bottles using the bottle's original label as a built-in "resist" to the etchant. makezine.com/go/etch

☼ Primer: EL Wire Louis M. Brill and Steve Boverie explain electroluminescent (EL) wire, aka "lightwire" — how it's built, how it works, and how it can be used in creative electronics projects to light up the night. makeprojects.com/Wiki/elwire

1+2+3 Cut-and-Fold Center Finder
By Andrew Lewis

Finding the center of a circle is easy when you have the right tool. This cut-and-fold cardboard center finder is ideal for all those fiddly measuring jobs.

1. Measure and mark.

On your 10" square of cardboard, use the ruler and pencil to mark a diagonal line from the bottom left corner to the top right corner. Mark a vertical line ½" from the left side, a horizontal line ½" from the bottom, and another horizontal line 1" from the bottom.

2. Cut.

Using your utility or craft knife, cut out a ½" square from the bottom left corner of the card, using your ½" pencil lines as a guide.

Next, cut a triangle out of the card. Starting at the top right corner, cut along the diagonal line until you reach the 1" horizontal line at bottom left, then cut along the 1" line to the right edge of the card.

3. Fold and tape.

Use your knife to lightly score along the ½" border lines so that the card will fold easily, then bend the left and bottom ½" borders 90° to form a corner. Tape the corner to hold the cardboard edges at 90°.

Use It

To mark the center of a circular object (for example, the top of a paint can), place your center-finding tool so that its folded edges touch the object's outside edges.

Now draw a line across the object, using the inside diagonal edge of your center finder as a guide. Rotate the object 90° and draw a second line. The point where the 2 lines intersect is the center of the circular object.

NOTE: The maximum diameter of the circular object is roughly equal to twice the center finder's outside edge length — so your 10" center finder can be used with objects up to about 20" in diameter. To make a larger center finder, just follow the same instructions but use a larger square of cardboard.

Andrew Lewis is a keen artificer and computer scientist with interests in 3D scanning, computational theory, algorithmics, and electronics. He is a relentless tinkerer who loves science, technology, and all things steampunk.

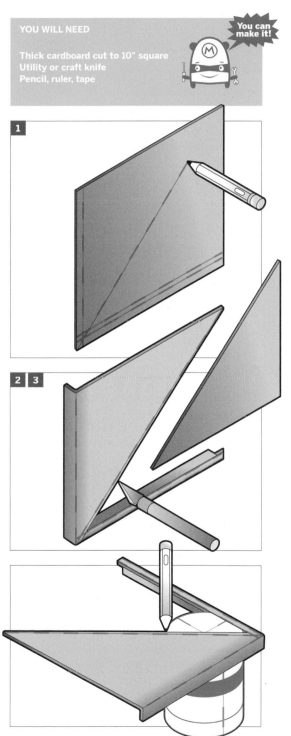

YOU WILL NEED

Thick cardboard cut to 10" square
Utility or craft knife
Pencil, ruler, tape

You can make it!

Cargo Bike Power

Car-free carrying makes a comeback. By Joshua Hart

If you want to wean yourself away from petroleum dependence, try a cargo bike. A good one lets a relatively fit adult transport 500 pounds of stuff across level ground, so a few bags of groceries are cake. In recent years, increasing concerns about the environment and energy dependence have made cargo bikes a hot topic for makers and tinkerers who want to design an important piece of the future.

The concept isn't new, of course. Bikes specially designed to transport cargo are nearly as old as the bicycle itself. Starting in the late 19th century, cargo bikes were used in many developed countries to carry items that would otherwise have required horsepower, and they continued to be used to deliver things like bread, milk, and mail until after World War II, when cars and trucks took over in industrialized countries.

But designing cargo capacity into pedal-powered transportation is a challenge that not only presents questions of where, what, how, and how much to carry, it also complicates all the usual bike design tradeoffs — balancing weight, strength, cost, gearing, frame geometry, wheel size, and so on.

It's a vast, messy, multivariable problem that can't be optimized for all applications. But over the generations, a few popular strategies have evolved that do a good job with common types of cargo and terrain. Makers like Joshua Muir and Saul Griffith are adding their own ideas to the field, building a revolution not just in cargo-bike design, but in the way people live.

Classic Carriers

How do people carry loads on bikes? Many popular "cargo bikes" are just regular bikes with cargo

Photograph by Paul Gower

capacity added on. Good old-fashioned *panniers* and *bike trailers*, for example, come in various shapes and sizes, and you can detach them to regain your bike's performance. But panniers have limited capacity, and trailers can be difficult to maneuver — tip-prone energy drags, rolling behind and clattering around.

To make room for loads on the bike itself, you can shrink the bike's front wheel and fork, lengthen the head tube, and attach a basket in front. This is the *tradesman's bike*, aka *deli bike*, *butcher's bike*, *post bike,* or *porteur*. With the basket attached to the frame, rather than the handlebars, steering is unaffected. One model, the Pashley Mailstar, is still used to deliver mail in the U.K., although they're sadly being phased out in favor of vans.

For even more capacity, you can move the front wheel forward, connect it back to the handlebars via a pair of curved steering rods, and carry the cargo low (and therefore stable) in between.

Such was the principle behind the Belgian keg bike (now commonly called *long john* or *long haul*), originally designed for a land of great beers and flat topography. Popular with messengers, these bikes carry heavy cargo (comparable to a

2-foot by 1½-foot, 160-pound beer keg) where the driver can easily monitor it, which especially helps riders carrying living "cargo" like kids and pets.

Inspired by this classic design, Portland, Ore., cargo-bike maker Metrofiets created the Beer Bike for a local brewery, which carries two kegs (with taps made from bicycle parts), an inlaid wood bar, a pizza rack, and a wood-paneled sound system.

Tadpoles, aka *Christiania trikes* (named after the car-free district in Copenhagen where many are made), are sturdy cargo tricycles that carry large loads, including children, between two front wheels. Like other tricycles, they're less maneuverable than bikes — wider and prone to tipping during turns.

A different cargo-trike approach is to put the two wheels and cargo in back, as with recumbent tricycles, where the driver sits low with legs extended forward.

Life After the Xtracycle

Perhaps no other U.S. company has done more to promote the potential of bicycles to transport cargo than Xtracycle (xtracycle.com), started by mechanical engineer Ross Evans.

In the spring of 1995, Evans began studying bicycle use in the developing world. He noticed, particularly in Central America, that while conventional bikes were abundant, they weren't as useful to people as they could be.

He set about designing an inexpensive cargo carrier that was lightweight, maneuverable, stable, and able to travel down narrow paths. This led him to develop the "longtail" frame-extension kit, a bolt-on attachment that extends the length (and cargo-carrying capacity) of regular bicycles, sold by Xtracycle under the name FreeRadical.

The kit attaches just behind the bike's bottom bracket and clamps to the rear-axle drop-outs, moving the rear wheel back about 15 inches (and swapping in a longer chain). The frame extends back to hold the rear wheel and wraps around behind it to support four upright tubes that carry a variety of accessories, like racks and decks for holding child seats or large panniers.

Many of today's cargo-bike designers were inspired by Evans. The explosive cult popularity of the smooth-riding Xtracycles has inspired a subindustry of compatible add-ons that take the cargo bikes in different directions — like the Stokemonkey electric motor assist kit (*MAKE Volume 11, page 82*), which lets a rider haul hundreds of pounds up steep hills that would be impossible for anyone but Lance Armstrong to pedal otherwise. You can pedal normally without any motor

resistance, but you can't use the motor without pedaling. As manufacturer Clever Cycles explains, "We don't believe in replacing human power with electricity; we believe in replacing cars."

But as with the Belgian keg bike, longtail bikes aren't all work and no play. The pedal-powered Margarita blenders and bike-based sound systems of Rock the Bike (*Volume 11, page 76*) were all originally built on Xtracycles.

Joshua Muir's Small Haul

Frame builder Joshua Muir (francescycles.com), who co-founded the influential Bike Church Tool Cooperative in Santa Cruz, Calif., loves to go bicycle camping with his 60-pound black Labrador, Soupy. In Belgium, a long john might enable a trip like this, but the steep roads of the Santa Cruz Mountains require something lighter.

So Muir designed a high-performance cargo bike, the Small Haul, which takes the classic long-john design as a departure point. In place of its rectangular cargo area, the Small Haul has an oblong, stretched-fabric load basket supported by thin, integrated frame tubes that curve out gracefully between the front wheel and the down tube. The resulting space frame looks more like part of a racing yacht than a traditional work bike.

Muir saved even more weight with the Small Haul's steering system. The classic long john includes idler steering, which uses lots of metal:

BACK AND FRONT: (Opposite) The Xtracycle FreeRadical frame extension installed on a bicycle, carrying an extended deck and panniers. (Above) Josh Muir's black Labrador, Soupy, on the road in the cargo basket of a Frances Cycles Small Haul.

an extended steering column connects down to two metal rods that run under the cargo basket then curve back up over the front fork. In addition to being heavy, the idler arms can get stuck when debris or cargo jam up the steering.

Instead, Muir adapted a less-common design based on sheathed cables, which bikes already use to transmit forces to brakes and derailleurs. The Small Haul's two steerer tubes, one under the handlebars and the other over the front fork, each attach to a perpendicular pulley. On each side of the bike, one cable end wraps backward around each of the identical pulleys, connecting the two.

The resulting system weighs less than a pound, handles like a regular bike, and is strong enough for Muir to haul Soupy, dog food, and camping supplies in front, plus other gear in panniers on the rear rack. The only nonstandard parts are the two aluminum pulleys. The Small Haul, including generator light, fender, rack, and pump, weighs approximately 37 pounds. (Muir also builds a heavier Cycletruck, which follows a more traditional idler-steering design and can carry 200 pounds.)

Unlike many custom bike builders (and all major manufacturers), Muir minimizes the energy and toxic materials he uses, sourcing components locally whenever possible. "I don't own a car, and drive little," he writes. "Most all of the materials I use are sourced within the U.S., and some come from the Bay Area. I build framesets by hand, one at a time, slowly."

Saul Griffith's Cargo Trike

Since the birth of his son, Huxley, in April 2009, regular MAKE contributor Saul Griffith has been designing and building cargo bikes. His aim, he says, is to create vehicles that can haul groceries and kids up steep San Francisco hills, eliminating the need for his family to use a car locally. With a fall 2010 launch date, Griffith's line of bikes will be sold as custom builds through his company Onya Cycles (onyacycles.com). If demand for any of the models takes off, he plans to start mass-producing them to sell at much lower prices.

Like Muir, Griffith has studied the past, including the bike-design "bible," Archibald Sharp's *Bicycles and Tricycles*, first published in 1896 (now printed by Dover Publications). But where Muir draws on his years of mastery, handcrafting lugged, steel-tube bicycle frames, Griffith takes a higher-tech

Maker

TYKE LIKE TRIKE: Saul Griffith carries his young son, Huck, in an early prototype of his cargo trike. The triangular wheelbase and innovative tilt-steering system ensure that the trike stays safely upright while it's stopped, traveling in a straight line, or turning at any normal radius and speed.

approach. For example, he uses CAD to model every piece of a bike, from the frame and wheels to the smallest component, trying them out virtually to see how they fit together. "Someday," he predicts, "everyone will design bikes like this, and there will be a new kind of 'digital artisan.'"

Griffith's line features three types of bikes, all of which he designed to replace cars for different uses: a light-hauling "runabout" that can be carried up stairs; a family-oriented longtail, where kids can ride in back; and a tadpole cargo trike for the heaviest hauling.

All three models use BMX wheels, which Griffith favors because they're cheaper and stronger than 27-inch or 700mm wheels, and they facilitate greater stability by keeping the center of gravity lower. (Smaller wheels also present marginally greater rolling resistance, but smooth tire tread and proper inflation matter more.)

The three bikes also have electric motor assist from an internal-hub motor with a planetary gear that can be set to different gearing ratios. Power comes from a lithium polymer battery and provides a range of 10 to 40 miles, depending on weight and terrain.

Of Griffith's three designs, the cargo trike may be the most revolutionary. Traditional tricycles are naturally stable while standing still or going in a straight line but are notorious for tipping when you turn, with even greater danger if they're motorized and carrying a heavy load. To remedy this instability, bike designers since Archibald Sharp's day have created tilt-steer systems that let wheels lean into the turn, although these have never appeared in a mass-market trike.

Most two-wheel steering systems use Ackerman steering, originally invented for horse-drawn carriages, in which the wheels connect to angled steering arms linked together by a tie rod. This allows each wheel to turn at a different radius around the same point, reducing friction and energy loss from tire slippage.

For Griffith's tilt-steer system, he extended the Ackerman geometry into three dimensions so that each wheel tilts as well as turns at its own angle. This required some "gnarly geometry," he says; to arrive at his final design, Griffith wrote and ran a 7,000-line simulation program that modeled all of the system's basic elements, analyzed the energy loss under any possible set of their dimensions

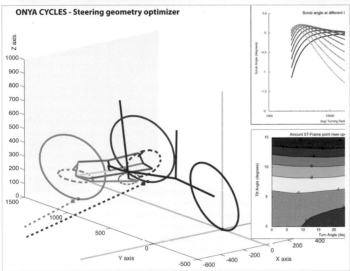

ONYA CYCLES - Steering geometry optimizer

GNARLY MATH: (Above) The tie rods and steering arms of Griffith's cargo trike extend standard Ackerman steering into three dimensions. This lets the trike's two front wheels lean as well as turn at different angles as they follow their own arcs, minimizing side-slippage and other sources of friction. (Left) Graph generated by the 7,000-line computer program that Griffith wrote to find the optimal geometry for his tilt-steering system.

and angles, and found the optimal combination, assuming a typical range of steering radii.

The result, converted from MATLAB to metal, is a tricycle that feels uncannily like a bicycle, with a one-minute learning curve that teaches riders to ignore the sight of a big cargo basket swaying side-to-side in front of them and to just ride normally.

Cargobiketopia Rising

As a transit planner and bike activist, I'm thrilled about what Muir and Griffith are doing, as well as many others who are designing not just new bikes but also local bike-sharing systems, maintenance collectives, and plans for reclaiming bike-friendly roadway and path infrastructure (often located along old rail rights-of-way, canals, or rivers).

Building level, regional bike routes has so far been carried out piecemeal, and mostly in the name of recreation. But bike routes could also become an arterial network for transporting people and cargo throughout re-localized communities. Combine this with "intermodal" connections to express buses and trains, and you've got a sustainable transportation system with quality of life as the driving force.

Such a system would knit communities together, rather than drive them apart. The able-bodied could transport cargo and other people around, and everyone could live, breathe, and congregate in safety, moving around without a wall of armor surrounding them.

Many people seem to be yearning for a new kind of transportation system, and whatever the future holds, the freight-carrying bicycle will be an important part of it. It's cheaper, more fun, healthier, simpler, more elegant, and more conducive to community than the alternatives. And designing, making, and pedaling new variations of cargo bikes around our cities and towns should keep us busy for many years to come.

Thanks to Stephen Bilenky and Erik Zo for helpful background information.

Joshua Hart (joshuanoahhart@gmail.com) is a Bay Area resident with 15 years' experience advocating for self-propelled transportation. His master's thesis, "Driven to Excess," was covered by more than 100 media outlets worldwide. He maintains a blog about transportation and climate issues at onthelevelblog.com.

Growing Bicycles

DIYers come from as far away as the United Kingdom to make bamboo bikes in this Brooklyn studio. By Jon Kalish

On a cold weekday morning, two of the three young men who run the Bamboo Bike Studio (bamboobikestudio.com) leave the Red Hook neighborhood of Brooklyn, N.Y., and drive a battered Toyota to New Jersey.

Bicycle makers Sean Murray, 27, and Justin Aguinaldo, 26, are embarking on one of their periodic bamboo harvests in the small sedan that belongs to the third member of the studio, Marty Odlin, who can't make the trip. Odlin has a day job managing the sustainable engineering laboratory at Columbia University in Manhattan.

Aguinaldo and Murray carry just two tools: a caliper, to measure the thickness of the bamboo stalks, and a small Japanese pull saw. They got a tip about bamboo growing wild on the grounds of

a nursery in New Brunswick, N.J., so they drive there and ask an employee if they can cut some down. They're told to help themselves.

Aguinaldo, a short, earnest cyclist who grew up in Fort Bragg, Calif., uses the caliper to tap the bamboo before it's cut. He gets a sense of the plant's density from the sound.

"If the bamboo's too watery, it's not as dense and not as strong," he explains. "It's harder to find the stuff that's denser, that's better for bikes that are ridden harder." Aguinaldo knows about riding hard. He and his two compatriots have logged thousands of miles on their bamboo bikes, mostly on New York City's potholed thoroughfares.

Murray, a former schoolteacher who declares on his outgoing voicemail greeting that he's living

THE BAMBOO WAY: Bike enthusiasts take weekend workshops at the Bamboo Bike Studio to build their own bikes. (Opposite) Frames are masked with painter's tape and secured in aluminum jigs while the joints are wrapped in carbon fiber soaked in epoxy. (Above) Studio co-owner Justin Aguinaldo tests out a handmade bamboo ride.

the dream of making bikes with his friends, has taken to trolling online gardening forums for leads on homeowners who are grappling with a bamboo invasion.

"One story I've heard a lot is, 'I got bamboo a few years back as a decorative plant and I still like the bamboo, but it's started to crawl into my neighbors' yards,'" he says while cutting down a 2-inch-thick stalk with his pull saw. "There's a kind of urgency brought on by the protests of the neighbors, you know."

The two bike builders harvest a species of bamboo known as *Phyllostachys angusta* that is common in the tri-state area of New York, New Jersey, and Connecticut. After a couple of hours cutting 3- and 5-foot lengths, they schlep the freshly cut green bamboo stalks in long canvas bags back to the car and fill up the trunk with the fruits of their harvest.

Aguinaldo and Murray return to Red Hook, a mostly low-rise neighborhood near Brooklyn's industrial waterfront, and carry the bamboo stalks into an old brick building with high ceilings. Everything in the long, narrow room that serves as the bike-building studio is homemade, including the

head-high aluminum frame-holding jigs and the oven used to dry out the bamboo.

For the drying process, they poke thin metal rods through the nodes inside the bamboo so it will dry out evenly when baked. A propane torch is used to cook and harden the skin of the bamboo, which turns from green to a beautiful tan. Then it's put into the oven for several hours at a low temperature.

Anywhere from two to six people make bikes during the weekend workshops run by Murray, Aguinaldo, and Odlin. It takes two long days and costs $932 to build your own bamboo bike. Bamboo, construction materials, and all bicycle components, such as wheels, handlebars, brakes, etc., are included in the cost. You can build just the frame for $632.

DIYers have come from as far away as California and the U.K. to make bamboo bikes in the Brooklyn studio. On a windy afternoon last November, Aguinaldo and Murray returned from a harvest and were surprised by a visit from Alexis Mills, a 29-year-old bicycle messenger who lives in Ottawa, Ontario.

He made a bamboo bike last October, as did his mother, Christina Mills, a 61-year-old doctor in

Maker

HAULING GRASS: (Clockwise from top left) The prepared bamboo, oven-cured, torch-hardened, and ready to make frames; a bamboo bike rack by Boo Bicycles; the personal ride of a Bamboo Bike Studio co-owner; and a close-up of its hardware connecting the bamboo chain stays, seat stays, seat tube, and down tube.

Waterloo, Ontario, who readily admits being one of those "tread lightly on the Earth" types.

"I just love the whole concept of making your own transportation," says Christina, who doesn't own a car but manages to get around pretty well in Waterloo on her four bicycles.

The first day of bike building is devoted to making the frame by connecting the bamboo with epoxy-soaked carbon fiber that looks like thin black ribbon. The bamboo-bike makers refer to this as "weaving the lug." After the epoxy hardens, the joints are hand-filed to smooth them out. At first glance, the finished joints look like they've been wrapped in black electrical tape. On the second day the bike components are attached to the frame, also with epoxy.

Early last December, two people in their 40s, both self-described tinkerers, made bikes in the Red Hook studio. Sari Harris, an information architect who designs interfaces for mobile phone apps, and David Anderson, a lighting technician who works on television and movie productions, were filing away when I dropped by.

Harris wanted a new bike because hers was more than 20 years old. She admits that, going into

the bike-building sessions, her mechanical skills were limited to changing a tire.

"Part of me is, 'Wow, I can make the frame,' and because I'll put all the components on, I'll learn a lot about the mechanics of how a bike works and maybe learn how to tune up my own bike," Harris says.

Anderson, who rides his bike all over New York City because his work takes him to a new set week after week, marveled at the bamboo he saw growing in Laos on one of his vacations there. The thing he likes about the Bamboo Bike Studio is that, "These guys are not a bike factory here. They're producing a way of making bikes, rather than producing bikes." The studio has no immediate plans to make and sell bikes, though Odlin does not rule it out at some point in the future.

Odlin, who is 28 and an accomplished skier, estimates that between the test bikes he and his partners have built and those made by DIYers who come for the weekend workshop, about 180 bikes have been constructed since the Bamboo Bike Studio began in January 2009.

Each weekday Odlin pedals his bamboo bike 12 miles over the Brooklyn Bridge and along the

Photography by Alan Esner; Boo Bicycles (bike rack)

Hudson River Park bike path to Columbia University on Manhattan's Upper West Side. He has been plagued by something that all people who ride bamboo bikes have come to endure — a constant barrage of questions about that bike.

"I ride with headphones even though I don't listen to music while I'm riding, so I can ignore people when they try to talk to me about my bike. If I talked to everybody who asked me about my bike, I'd never get to work," Odlin explains.

Ditto for Murray and Aguinaldo. Aguinaldo uses his bike for his business, the Mess Kollective, a bike messenger collective that has no office and is run entirely on iPhones.

Murray is soon relocating to the Bay Area, where he'll be setting up a San Francisco-based Bamboo Bike Studio, with weekend workshops already scheduled into the new year. The three founders also "think globally," with a portion of all class fees going toward efforts to seed the first bamboo bike factory in Ghana.

This past summer the Bamboo Bike Studio started selling kits for DIYers who want to build a bike frame at home. The kits include a jig, some tools, epoxy, carbon, and a limited number of metal parts, such as the special dropouts for the rear wheels. They cost less than $500 — bamboo costs extra, although the studio plans to crowdsource a harvest map for those who want to find local bamboo.

There are at least three detailed how-tos for making a bamboo bicycle on instructables.com. The safety of DIY bamboo bikes has been questioned by Calfee Design's Craig Calfee (calfeedesign.com), a high-end bike maker in La Selva Beach, Calif., and a pioneer in using bamboo for bicycles.

Calfee, who developed the technique of wrapping epoxy-soaked fiber around bamboo junctions in 1995, told me that building a bamboo bike using "the wrong techniques" could result in serious injury. But he says he assumes the bikes made by Odlin, Murray, and Aguinaldo in Brooklyn are structurally sound.

"I'm more concerned with the average DIYer," says Calfee. "It's possible to build a bamboo bike that rides just fine soon after it's completed. But after the bamboo ages or the resin shrinks, the bamboo can separate from the wrappings, causing very unexpected results."

Bamboo bicycles may seem like the ultimate mode of environment-friendly transportation, but if you buy one as opposed to making one yourself, they can cost a whole lot of green.

Calfee Design's bamboo bike frames, which have joints made from epoxy-soaked hemp, sell for $2,695 and $3,195; but he also started a company called Bamboosero, which imports bamboo bike frames made in Africa and sells them starting at around $700. Models include mountain, cargo, city, and road bikes.

In Portland, Ore., **Renovo Hardwood Bicycles** (renovobikes.com) sells laminated bamboo bike frames starting at $1,495–$2,650, plus extra for full builds.

There are two bamboo bike makers in Fort Collins, Colo. **Panda Bicycles** (pandabicycles.com) makes bikes with bamboo "tubing" connected using a proprietary steel-joint design. The company offers three models ranging from $1,600–$2,150 for frame only, and $2,100–$3,250 for full builds.

Boo Bicycles (boobicycles.com), also in Fort Collins, was started in 2009 by Nick Frey, a 23-year-old pro cyclist and mechanical engineer. His bamboo bikes, which boast carbon fiber joints, are handcrafted by James Wolf, an American furniture maker who lives in Vietnam. Boo sells five models, with frames ranging from $2,625–$2,985, plus customs.

Organic Bikes (organicbikes.com), which is owned by the Wisconsin retailer Wheel and Sprocket, sells a bamboo bicycle called the Dylan for as little as $1,000. It's made from compressed bamboo dowels connected by recycled aluminum lugs.

A Danish bike maker, **Biomega** (biomega.dk), also uses aluminum lugs on its bamboo bike, which was developed by award-winning industrial designer Ross Lovegrove with the expertise of Brazilian bamboo specialist Flavio Deslandes.

With all these companies jumping on the bamboo bandwagon, the guys in Brooklyn are concerned that bamboo bikes might become a fad that eventually dies out.

"We feel like we're building something with more enduring value than that," says Odlin. "Everyone who leaves the studio says, 'Wow, my bike is my favorite object now.' They have such a connection to this thing that came together under their own hands. They may not come here to have that connection to their bicycle, but that's what they leave with. Everyone leaves with that."

Jon Kalish (jonkalish@earthlink.net) is a Manhattan-based radio reporter and podcast producer. He covers the DIY scene for NPR.

1+2+3 Ping Ponger
By Edwin Wise

The Ping Ponger uses almost half of a rubber racquetball as a disc spring that's *bistable* (it can be at rest in 2 possible states) to propel a ping-pong ball from a compact PVC launcher.

1. Make the PVC parts.

Cut all PVC pipe pieces to length. For the body, cut the ends off the snap tee so it's 2½" long, then press-fit the female adapter into the bottom of the tee and cut it off flush.

Glue the backstop rings together so that one edge is flush.

Bevel the inside edges of the barrel pieces with a file, so they curve to match the shape of the racquetball. Glue the 2 barrel pieces together.

For the handle, sand the 1" repair coupling so it fits into the 1½" pipe and glue it in place, then glue the tip of the 1" pipe into this coupling.

2. Make a disc spring.

Seat a racquetball into the beveled end of the barrel. Trace a line around the ball (parallel to the seam) and cut on this line, leaving a dome. This is your disc spring.

3. Put it all together.

Glue the backstop into the body (the snap tee), flush with the back end.

Set the rubber spring on the backstop inside the body. Slip the barrel in place over the spring, leaving just enough room for it to spring forward and back. Don't glue the barrel; you can remove it to change the spring.

Slip (don't glue) the handle into the bottom of the body. The handle also acts as a ping-pong ball holder and ramrod.

Now Ping-Pong Away!

Put a strong, name-brand ping-pong ball into the barrel, and push it into the spring using the handle, until the spring sticks open (back), gripping the ball.

Through the backstop, poke the tensed back of the spring to make it un-spring. Pong!

Edwin Wise is a software engineer and rogue technologist with more than 25 years of professional experience, developing software during the day and exploring the edges of mad science at night. simreal.com

YOU WILL NEED

Rubber racquetball
Ping-pong balls
Marker, knife, saw, and half-round file
PVC glue, epoxy, and sandpaper

PVC pipe and fittings:
For the handle:
1" pipe, 1½" length
1" repair coupling
1½" pipe, 4½" length
For the body:
2" snap tee **part #463-020S from** flexpvc.com
1½" female adapter coupling

For the backstop:
2" pipe, 1" length
⅝"-wide ring from the insert coupling (below)
For the barrel:
2" pipe, 1¾" length
2" insert coupling **ProPlumber model #PPFC200, Lowe's part #153807 (as used in the Boom Stick, MAKE Volume 13). Cut it down to 2" long. Also cut a ⅝" ring from the scrap.**

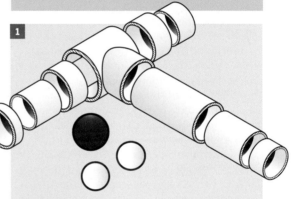

Photograph by Ed Troxell

Make: SPACE

The opportunities to explore space have never been greater, thanks to new, inexpensive technologies and NASA's realization that amateur enthusiasts and small companies are its greatest resource.

In the following pages, we'll show you how to get involved in the next great era of DIY space science. The countdown begins now!

>>

Making Your Own Satellites

Build and launch your own sat for as little as $8,000.
BY CHRIS BOSHUIZEN

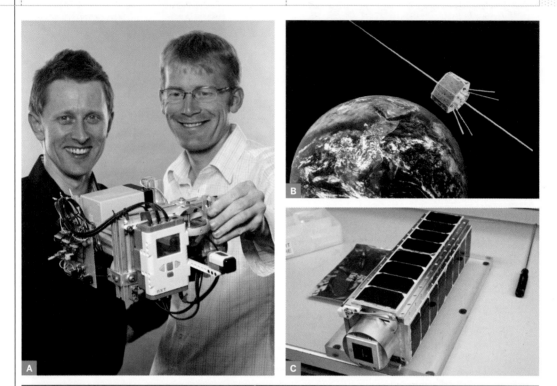

AMATEUR NIGHT: Fig. A: With a Lego NXT system and $500 in parts, a team of International Space University students built this fully functioning prototype satellite, shown by NASA's Chris Boshuizen (left) and Will Marshall. **Fig. B:** AMSAT-OSCAR 7 has been orbiting Earth since 1974. **Fig. C:** PharmaSat, developed by NASA Ames Research Center and launched in 2009, is a three-unit CubeSat to study antifungal drugs in microgravity.

It's often said that there's nothing you can't make at home, and even the final frontier is not too remote from the hands of a well-equipped group of DIYers.

Amateur groups have been launching their own satellites into space for 40 years. Today, cheap technologies and novel launch strategies are helping DIYers build and launch more satellites than ever.

The most successful amateur satellite is AMSAT-OSCAR 7 (Figure B), which has been in orbit for 36 years and remains semi-operational to this day. Originally launched as an experiment on new types of transceivers, it's still able to send, receive, store, and forward messages.

Let's take a look at the present and future of amateur satellite building.

AMSAT and the OSCAR Spacecraft

AMSAT (the Radio Amateur Satellite Corporation) is a group of organizations that design, build, and operate amateur satellites. AMSAT organizations from 23 countries have launched their own "orbiting satellites carrying amateur radio" (OSCARs for short). The first of these was launched in 1961, a mere four years after the launch of *Sputnik 1*.

AMSAT spacecraft range in size from 2kg up to 50kg and have become more sophisticated over the years. To date, most have been placed in orbit around the Earth, but AMSAT even has guidelines

Photography by Garry McLeod (A); NASA (B, C)

for spacecraft that can travel to other places in the solar system. AMSAT-DL in Germany is planning the GO-Mars/P5A spacecraft to be launched to Mars.

AMSAT pioneered the launching of amateur satellites as secondary payloads on commercial rockets by filling the empty space around the larger satellite that was paying for the ride. While it's not always free, it's still a very cost-effective way of getting something to space.

Tubes and Cubes

The pioneering satellites of AMSAT were designed and built from scratch, but these days there are standardized nanosatellite designs and even starter kits. TubeSats and CubeSats are leading the way with standards for size and delivery to space.

CubeSats are based on a 10cm-cubed design originally developed for universities. They can be connected into units three cubes long, and they're launched from a rectangular "peapod," pushed out into space like peas squeezed from a pod.

CubeSat missions have carried a variety of high-tech gadgetry, including high-resolution cameras and even instruments to measure earthquakes. NASA recently launched its own CubeSat mission, PharmaSat (Figure C), to examine the effectiveness of antifungal drugs in space. NASA hopes to do more cost-effective microgravity research this way.

A CubeSat kit to build a functioning satellite could cost between $5,000 and $300,000, but you can build one much cheaper with some clever planning.

Similar to CubeSats, **TubeSats** are delivered from a long cylinder attached to the upper stage of the rocket. A TubeSat kit from **Interorbital Systems** costs only $8,000, including launch! According to the company's website, "Planet Earth has entered the age of the Personal Satellite." I want mine!

Lego and Smartphones

At NASA Ames Research Center we're trying to discover how cheaply we can build a spacecraft. With the **Lego Mindstorms NXT** system and about $500 in other parts, we built a fully functioning prototype satellite (Figure A).

We're also very interested in **smartphones**, which are bristling with sensors and have onboard computers more powerful than nearly every satellite ever put into space. In fact, a smartphone has nearly all the systems of a spacecraft except solar panels and propulsion! With a bit more work, we think we'll be launching the first cellphone-based satellite one day (see page 74).

Getting Up There

While you might be able to build a functioning satellite for very little money, there are still two problems to overcome. The first is finding a launch on some kind of rocket. And if you end up sharing a ride, the next problem is convincing the operator that your satellite is not a danger to the much more expensive one they're putting on the rocket.

To reduce the risk of damaging either the rocket or the other spacecraft, your satellite needs to go through extensive testing to ensure that it's safe. This will include putting it through extreme hot and cold cycles, subjecting it to simulated vacuum in a large vacuum chamber, and shaking it aggressively on a vibration table. If your satellite can survive this punishing regime and still function as you designed it, then it should also survive the rocket ride and operate in space for some time.

Getting a launch for your object is probably the most expensive part of doing anything in space. You could volunteer for an AMSAT project, and we've already mentioned TubeSats, which offer a complete delivery service. If you're building your satellite through a university or other educational group, NASA's **Project ELaNa** (Educational Launch of Nanosatellites) can get a CubeSat into space for around $30,000. **Andrew's Space** is a firm that brokers commercial launches for SpaceX's exciting Falcon 1 and Falcon 9 rockets. While these launches cost more than ELaNa, the rockets can carry much heavier satellites — anything from 1kg to 300kg!

If you want to try your satellite out but not necessarily have it in space for months, you might consider putting it on a short-duration **high-altitude sounding rocket**. Many amateur groups launch rockets to over 10km in altitude. Another option is a **high-altitude balloon** ride. Balloons are much more gentle, and can take your experimental satellite up to 30km without much hassle. You then have plenty of time to get it settled in, started up, and ready to do what it was designed for. And if dropping it was part of the plan, you should get at least 30 seconds of free-fall time from that height!

Spacebridge is an offshoot of the San Francisco Bay Area hackers group Noisebridge. They're developing their space voyages using balloon launches and smartphones, but hope to be soon launching their own rockets. And why couldn't you?

Chris Boshuizen is the small spacecraft technical liaison at NASA Ames Research Center. He has always wanted to be an astronaut.

Listening to Satellites

Tune in to space with a homemade yagi antenna.

BY DIANA ENG

One of my favorite things to do is talk with other ham radio operators through satellites or the International Space Station (ISS). To do this, I stand on a rooftop and tune a handheld multiband radio while tracing the orbit of a satellite or the ISS with my homemade yagi antenna.

Orbiting satellites such as AO-51, SO-50, and AO-27 act as repeaters, relaying signals from low-power transceivers like mine back to hams elsewhere on the planet. So if you know where to aim the antenna, you can communicate around the world via space.

The ISS also has a repeater, and occasionally, when we're lucky, the astronauts themselves exchange transmissions to communicate with hams on the ground.

To listen to these signals from space, you don't

Photography by Diana Eng

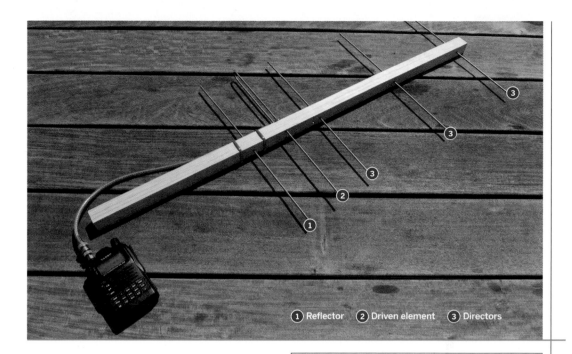

① Reflector ② Driven element ③ Directors

have to be a licensed ham radio operator, or even stand on the roof. You can do it in your own backyard with an off-the-shelf UHF FM radio. The whip antenna on the radio might let you hear satellites and the ISS, but you'll get far better reception by making your own yagi antenna, which takes about an hour and costs less than $25 (not including the cost of your radio) using materials from your local hardware store.

If you do have a ham radio license and a UHF/VHF transceiver, you can upgrade this antenna with VHF elements so that it can both send and receive transmissions.

A yagi antenna has three types of elements, consisting of metal rods of varying lengths and quantities. The *driven element* is a dipole antenna that's connected to the radio and receives the signal, just like a whip antenna. The *reflector* is positioned behind the driven element, where it acts as a mirror by bouncing signals from the satellite forward to the driven element. *Directors* are one or more rods that act like a lens, focusing the incoming signal onto the driven element. Both the reflector and the directors improve reception from whatever direction the antenna points.

The antenna design I use comes from Kent Britain's (WA5VJB) "Cheap Antennas for Low Earth Orbit" (available at wa5vjb.com/references.html), which is a great reference for building many different types of yagi antennas.

MATERIALS

UHF FM radio like a police scanner, such as the Uniden BC72XLT handheld scanner, amazon.com, $85. Or, if you have a ham radio license, a UHF transceiver such as the Yaesu VX-7R, universal-radio.com/catalog/ht/0777.html, $308.

Square wooden dowel, 1"×1"×30" or longer if you want a handle longer than 10" for attaching to a tripod or mounting the radio. Approximately $2–$3 at hardware or craft stores.

Brass rods, ⅛" diameter, 36" long (3) Uncoated brazing rods work, but almost any brass rod or tube will do as long as it's approximately ⅛" diameter; $3–$4 each at a hardware or craft store.

Coaxial cable, RG-58, with BNC connectors, 3' universal-radio.com/catalog/cable/cable.html, order #4616, $4

Nylon cable ties (2) aka zip ties

TOOLS

Hacksaw
Soldering iron and solder
Glue gun and glue
Wire strippers and cutters
Ruler
Marker
Vise
Drill, or drill press, and ⅛" drill bit
File, smooth cut, flat, approx. 10"
Wooden dowel or broomstick, approx. ¾" diameter
Computer with an internet connection
Pencil

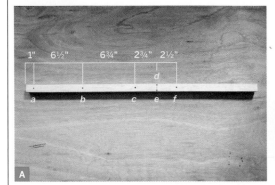

A

B

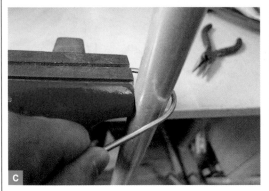

C

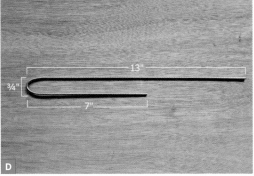

D

Build the Yagi Antenna

Time: 1–2 Hours

**Complexity: Building Antenna = Easy;
Receiving Signals = Medium**

1. Measure and cut.

Use a pencil and ruler to draw a centerline down one long side of the wooden beam. Then measure and mark hole locations on the centerline (except holes *d* and *e*) at the following intervals: hole *a* 1" from one end; hole *b* 6½" from *a*; hole *c* 6¾" from *b*; holes *d* and *e* 2¾" from *c*, ⅝" apart and equidistant from the centerline; and hole *f* 2½" from *d* and *e*.

Drill ⅛" holes completely through the beam at each point (Figure A). Be careful when drilling *d* and *e* not to blow out the sides of the beam.

Use a marker and ruler to mark 5 pieces of brass rod at the following lengths: 21", 13½", 12½", 12¼", and 11¾". Secure the rod in a vise, cut to the measured lengths using a hacksaw, and file the ends so they're no longer sharp and dangerous (Figure B).

To make the driven element, place the 21" rod in the vise, mark it 13" from one end, center the mark on the broomstick, then bend it 180° around so it's J-shaped (Figure C). Trim the rod so it measures 13" from one

end to the center of the ¾" curve, and 7" from the other end to the center of the curve (Figure D).

2. Assemble the parts.

Insert the 11¾" element into hole *a*, the 12¼" element into *b*, the 12½" element into *c*, the J-shaped (driven) element into *d* and *e*, and the 13½" element into *f*. Center all the elements, and secure them in place with hot glue (Figure E).

To prepare the coaxial cable, cut off one of its connectors and strip 3" of outer insulation off that end, being careful not to cut the wires. Separate the outer wires, twist them to one side, and strip 2" of insulation off the inner wire (Figure F).

Connect the coax cable to the 2 parts of the driven element near where they enter the wooden beam. Wrap the cable's inner wire around the short leg of the J, and the twisted outer wires around the long leg. Solder the wires in place (Figure G).

Secure the coax cable with a couple of zip ties (Figure H). Your antenna is done (Figure I)!

3. Receive signals from space.

To use your antenna, you need to find out where to point it and what frequency to tune in to. To find a good satellite target, visit heavens-above.com.

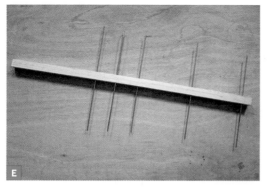

To specify your location, select a Configuration option (map, database, or manual), plug in the necessary info, then click Submit. From your new location-specific homepage, select "All passes of the ISS" to track the International Space Station or "Radio amateur satellites" to track a ham radio repeater satellite (Figure J, following page).

On the Radio Amateur Satellites page, click on one of the radio satellites you want to track from the Satellite column (such as AO-Echo, aka AO-51; SaudiSat 1C, aka SO-50; or AO-27), then show its pass chart by selecting "Passes (all)" above the globe (Figure K).

The pass chart lists all the satellite passes for the next few days. Each pass is listed by its times and locations in polar coordinates, for its start, maximum altitude, and end, with each pass typically taking about 10 minutes. The start and end points are defined as when the "bird" appears 10° above the horizon, and the maximum altitude (in degrees above the horizon) will vary. The azimuth for each location is listed in compass points (Figure L).

Make sure your location is listed correctly on the chart, and pick a pass during which the satellite will come close to directly overhead. Look for max altitudes that are 45° or higher — the higher, the

better. In the example here, the second pass, on July 17 at 3:50, looks good since its altitude reaches 75°, but the first pass, on July 16 at 16:55, only comes up to 18°, which is very close to the horizon and difficult to pick up.

Next, find the frequency to tune in to. Satellite repeaters work with 2 different frequencies — an uplink and a downlink. You listen to signals received via the downlink. (If you wish to transmit, you'll need to program in the uplink frequency as well.)

To find a radio satellite's current frequencies, you have to refer to the authoritative web page for each individual satellite. Some references online, including AMSAT (amsat.org), aggregate frequency information for multiple satellites, but these can be incorrect and you often need to dig deeper.

What you want is a current update or schedule with uplink and downlink frequencies, and this data

J (Heavens-Above screenshot)

Hosted by GSOC — DLR

Shuttle Missions
Next Shuttle mission will be STS-133, currently scheduled for November 1st.

Configuration
Current observing site: My Location, 33.6146°N, 80.9912°W
select from map or from database or edit manually
Registered user login | Why register?
Create new user account
AvantGo channel discontinued, please click here for details

Satellites
10 day predictions for: ISS | X-37B | Genesis-1 / 2 | Envisat | HST
Daily predictions for all satellites brighter than magnitude:
_____ 4.5 | 4.5 (dimmest)
All passes of the ISS including daylight and invisible passes.
Iridium Flares
next 24 hrs | next 7 days | previous 48 hrs
Daytime flares for 7 days – see satellites in broad daylight!
Solar System – where are they now?
Radio amateur satellites 24 hour predictions (all passes)
Satellite finder from database
Height of the ISS – how does it vary with time

Astronomy
Comets currently brighter than mag. 12
C/2009 R1 McNaught | 10P Tempel 2 | 2P Encke | C/2010 A1 Hill
43P Wolf-Harrington
Minor planets currently brighter than mag. 10
1 Ceres | 4 Vesta | 6 Hebe | 15 Eunomia | 2 Pallas | 8 Flora | 7 Iris | 29 Amphitrite | 3 Juno
Whole sky chart
_____ and Moon data for today
_____et summary data
_____et details (under construction)

Current position of tl

K (AO-27 Information screenshot)

| Home | Passes (visible) | Passes (all) | Orbit |

AO-27 - Information

Identification
USSPACECOM Catalog No.: 22825
International Designation Code: 1993-061-C

Satellite Details
Orbit: 788 x 800 km, 98.5°
Category: Amateur Radio
Country/Org. of Origin: USA
Mass: 12 kg
Dimensions: 150x150x150mm
Intrinsic brightness (Mag): 8.3 (at 1000km distance, 50% illuminated)
Maximum brightness (Mag): 7.3 (at perigee, 100% illuminated)

Launch
Date (UTC): September 26, 1993

For further information, please click here.

L (AO-27 All Passes screenshot)

HEAVENS ABOVE

AO-27 - All Passes | Home | Info. | Orbit | Prev. | Next |

Search Period Start: 00:00 Friday, 16 July, 2010
Search Period End: 00:00 Monday, 26 July, 2010
Observer's Location: My Location (33.6146°N, 80.9912°W)
Local Time: Central European Summer Time (GMT + 2:00)
Orbit: 788 x 800 km, 98.5° (Epoch 15 Jul)

Click on the date to get a ground track plot.

Date	Starts			Max. Altitude			Ends		
	Time	Alt.	Az.	Time	Alt.	Az.	Time	Alt.	Az.
16 Jul	16:55:31	10	SW	16:59:03	18	WSW	17:02:35	10	WNW
17 Jul	03:50:03	10	NNE	03:55:10	75	E	04:00:24	10	S
17 Jul	14:49:24	10	E	14:51:30	12	ENE	14:53:37	10	NE
17 Jul	16:25:54	10	S	16:30:47	40	WSW	16:35:40	10	NW
18 Jul	03:22:18	10	NE	03:26:55	33	ESE	03:31:34	10	SSE
18 Jul	05:03:26	10	NW	05:06:13	15	WNW	05:09:01	10	WSW
18 Jul	15:57:14	10	SSE	16:02:29	88	SW	16:07:45	10	NNW

---

is unfortunately not published in a standardized manner. With AO-51, for example, AMSAT's top-level listing links to a page that shows all the frequencies the sat is capable of, but not which ones are currently active. For that, you must click through to the AO-51 Control Team News page at amsat.org/amsat-new/echo/CTNews.php.

For HO-68, to give another example, you need to click the Organization listing to CAMSAT (camsat.cn), an amateur satellite organization in China, where you'll see the sat's active frequencies listed under its former name, XW-1. In a pinch, you can always just Google the satellite's name to find its authoritative source.

Once you've determined your target sat's current downlink frequency (example: 436.7950MHz FM), tune your radio to that frequency, and you're ready to go out and listen. Aim your yagi antenna directly at the satellite, with the shortest rods (directors) closest to the satellite and the longest rod (the reflector) farthest away. When the pass starts, aim the yagi toward the satellite (Figure M), then sweep it right and left slightly until you hear something. You can also move the antenna up and down slightly as you sweep right and left. Also try rotating the antenna by twisting your wrist, adjusting its polarity to receive a stronger signal.

If you're using a whip antenna, hold it perpendicular to the satellite, and keep it perpendicular while you rotate it to get a clearer signal (Figure N).

Trace the path of the satellite's orbit according to the pass chart, so that at its maximum altitude and its end time, the antenna is pointed in the corresponding locations. In our example, the antenna should be pointed east at 75° above the horizon at 3:55, and south at 10° above the horizon at 4:00. It can be difficult trying to catch the satellites, and you may spend a lot of time not hearing anything. The best method is to move the antenna around in small side-to-side and up-and-down motions until you hear a bit of audio.

The Doppler effect makes the frequency vary by 0.010MHz, so as you trace the satellite's path you'll also need to twiddle the tuning a bit. Add 0.010MHz to your target frequency early in the pass, then gradually dial it down until it's approximately 0.010MHz less than the listed downlink frequency by the end time.

The FM satellites repeat whatever they receive, so you'll hear whoever's signal is strongest. (Another type of satellite, linear transponders, can handle multiple conversations at once, but these are harder to use and require a more expensive single side-band, not FM, radio.)

Hamspeak

When you eavesdrop on ham radio satellites and ISS, you'll probably hear a lot of letters, numbers, and strange words, like "KC2UHB Foxtrot November three one … roger roger." One reason is that ham

operators use a phonetic alphabet to make themselves clear through the static and interference, so that "P" sounds nothing like "T," for example.

The ham ABCs are: Alpha, Bravo, Charlie, Delta, Echo, Foxtrot, Golf, Hotel, India, Juliet, Kilo, Lima, Mike, November, Oscar, Papa, Quebec, Romeo, Sierra, Tango, Uniform, Victor, Whiskey, X-ray, Yankee, and Zulu.

Also, orbits don't last very long, so radio operators extending their reach via satellites tend to communicate quickly, following the same general dialogue. Here's an example:

"Kilo Charlie two Uniform Hotel Bravo." (Hi, my call sign is KC2UHB, does anyone want to talk to me?) A call sign is like a screen name assigned to ham radio operators when they receive their license. Some operators have vanity call signs like NE1RD.

"KC2UHB from Whiskey two Victor Victor please copy Foxtrot November three one." (I hear you KC2UHB, my call sign is W2VV and I am in Maidenhead location FN31.) The Maidenhead system divides the Earth into grid squares as shorthand to describe locations, and FN31 covers most of Connecticut and some of New York State. You can look up grid square locations online at levinecentral. com/ham/grid_square.php and elsewhere.

"W2VV, QSL this is KC2UHB, Echo Mike eight nine." (W2VV, I received your transmission, my loca-

tion is EM89.) KC2UHB is in central Ohio.

"QSL. Thank you for the contact. 73." (I received your transmission. Thank you for the contact. Goodbye.)

"73." (Goodbye.)

Just as we text each other abbreviations like OMG, BRB, TTYL, LOL, BF, GF, and <3, ham operators have their own, much older shorthand that was originally based on Morse code but became spoken with the advent of voice transmissions — much like when people say "Oh em gee" or "Be eff eff" today. Here are some ham abbreviations you may hear:

73 = goodbye, best wishes
88 = xoxo
OM (old man) = a friendly term for a male ham, a boyfriend/husband if described as "my OM"
YL (young lady) = a female ham, a girlfriend if described as "my YL"
XYL = wife
QSL = confirmation of message received
QRP = operating with low power
HT (handy talky) = a walkie-talkie

▐◀ To learn more about how a yagi antenna works, watch Diana Eng's MAKE video on directional antennas, aka "Seeing Radio Waves With a Light Bulb," at makezine.com/go/yagi.

Diana Eng (dianaeng.com) is a fashion designer who works with technology, math, and science. She is author of *Fashion Geek: Clothes, Accessories, Tech* (North Light Books, 2009) and is the ham radio correspondent for Make: Online (makezine.com).

Weather Balloon Space Probes

Sense, signal, and snap photos in the stratosphere.

BY JOHN BAICHTAL | ILLUSTRATION BY JAMES PROVOST

A bunch of hackers drive into the desert with a trunk full of equipment: a weather balloon, a tank of helium, and a styrofoam cooler loaded with cameras and sensors. After filling the balloon, they release it and watch it hurtle skyward, the cooler and a parachute dangling beneath.

The hackers track the balloon on laptops as it rises to black-sky altitudes 20 miles up, whereupon the balloon bursts and the payload floats down. Team members on dirt bikes race to recover the package, checking their mobile phones for SMS texts containing GPS coordinates of the landing site.

While this sounds like a scene from some hacker novel, launching and recovering near-space balloon probes is easier than ever, and dozens of amateur groups — ham radio enthusiasts and hackers alike — are doing it now.

E Cut-Down

Legal flights require a cut-down mechanism to separate the balloon from its payload and parachute after a set time or in response to a signal. One simple cut-down circuit uses a relay to discharge a dedicated 9-volt battery through a high-resistance Nichrome wire coil that's wrapped around a nylon cord. Close the relay, and the coil melts the cord.

F Enclosures

An enclosure protects payloads from the extreme temperatures of the upper atmosphere and the impact of hitting the ground. Most amateurs use a foam cooler or construct an enclosure out of extruded polystyrene (XPS), which costs a pittance and doesn't crumble. You can also use small Pelican cases (pelican.com, prices vary) to protect individual devices, but this adds weight. A coat of day-glo paint will make enclosures easier to spot; be sure to write your phone number prominently on the outside of all enclosures.

A Balloons

Most groups choose the Kaymont 1,500g sounding balloon (kaymont.com, $105). It's relatively inexpensive, can lift 3.8kg, and is rated for a bursting altitude of 34.2km, or more than 110,000 feet. (It's possible, but not recommended, to eke out more lift by overinflating your balloon.) For helium, rent a tank from a local industrial/medical gas supplier.

B Radar Reflector

To make your balloon more visible to other aircraft, you may want to dangle a radar reflector from it, like the lightweight Emergency Radar Reflector from Davis Instruments (makezine.com/go/davis, $30).

C Parachutes

A typical choice is a 5' parachute rated for 4.7lbs from Rocketman (the-rocketman.com/recovery.html, $50). If in doubt, order the next size up — most chutes let you change the canopy size by adjusting the shroud lines. For the cord, you can use mil-spec nylon paracord rated for 95lbs. A 100' spool, more than enough for any balloon project, should cost less than $10.

D Instrumentation

Common instrumentation includes still and video cameras, and devices for measuring and recording humidity, altitude (or air pressure), temperature, acceleration, and magnetic field. For power, you'll want lithium-ion batteries rated to –40°C; you can test devices inside a cooler with dry ice and fans. When choosing instrumentation, make sure your total payload will weigh less than what the balloon can lift. You can use the free Canon Hack Development Kit (chdk.wikia.com) to control PowerShot cameras, or the lightweight BalloonSat Mini controller (nearsys.com, $19) to operate three sensors and a camera. Both are programmable in BASIC.

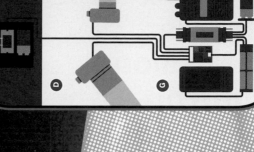

M4K3 24
707-827-7273

Make:
makezine.com

G Telemetry

You may be able to eyeball your probe until it lands, but a balloon can travel many miles, particularly in strong winds. Here are three popular lightweight tracking solutions:

The OpenTracker+ kit (argentdata.com, $32), which interfaces with APRS (aprs.org), the tracking system used by hams

G1 Android smartphone with the Icarus app (noisebridge.net/wiki/icarus), which collects GPS info and sends the coordinates in a text message every 60 seconds

Spot personal tracker (findmespot.com, $200, including one year of service)

Recovery

Once the payload deploys, you have to find it — even if it's miles away from the launch site. Terrain and roads permitting, you might be able to follow the flight with chase cars or motorcycles. Otherwise, you'll have to rely on tracking, good luck, and the kindness of strangers.

Regulations

Be sure to follow FAA regulations regarding launch sites and notifications, payload weight and density, and cord strength and cut-down mechanisms. Refer to the Society for Amateur Scientists guide at makezine.com/go/balloonfaa

+ For other resources, including additional references, trajectory predictors, FAA contacts, and a list of regional balloon launch groups and recent launches, see makezine.com/24/weatherballoons

Makers at Mission Control

Meet the elite NASA team that figures out how to fix the space station when things go wrong.

BY RACHEL HOBSON

WHEN IT'S HOUSTON'S PROBLEM: Pam Martin sits on console in Mission Control at NASA's Johnson Space Center, where she and other flight controllers guide astronauts through routine and emergency maintenance onboard the International Space Station.

Imagine trying to walk someone through the steps of changing the brakes on their car — over the phone. Now, imagine they're in outer space.

For Pam Martin and her colleagues at the Johnson Space Center in Houston, this is just another day at the office, only instead of helping someone fix a simple part on a car, she's helping astronauts repair and maintain delicate systems on the International Space Station (ISS), which is orbiting more than 200 miles above Earth and traveling at a whopping 17,500 miles per hour.

"A car is a very complex system, and we're talking about a massive space station so it's even more difficult," Martin says. "And you can't really talk on the phone all the time because you have communication gaps. If you're lucky, you might have a video camera that gives you some pictures every now and then."

Operations Support Officer Flight Controllers (OSOs) are in charge of training space station crews

Photograph by Thomas F. Murray

on daily maintenance procedures like changing out filters and making sure the toilet is functioning properly, as well as in specialized procedures for installing and activating new modules of the Station. OSOs are on console at ISS Mission Control, providing technical support, and are on call for emergencies onboard the ISS. On the Space Shuttle side of Mission Control are In-Flight Maintenance Flight Directors (IFMs), who provide similar training and support to the shuttle crews.

The OSO and IFM teams have their roots in the Apollo program, specifically Apollo 13, when flight controllers had to quickly come up with solutions for modifying carbon dioxide scrubbers in the Lunar Module when the Service Module was damaged by an oxygen tank explosion and crew members had to abandon the Command Module to take refuge in the Lunar Module. Using only materials the astronauts were known to have on hand, flight controllers fashioned a modified scrubber and then radioed the build instructions up to the crew. Materials included a sock and duct tape.

In 2007, Martin and the other OSO flight controllers had to come up with a solution to repair a torn solar array on the ISS and devised, in just a few days, a fix that would normally have taken months of planning and procedure writing. Using wire known to be in storage onboard the ISS, they devised a set of "cufflinks" to mend the tear in the solar array. Martin says the experience was intense but rewarding.

"It's like Apollo 13," she says. "We sit down and say, 'OK, this didn't work. It wasn't designed right — so what do we have onboard that we can fix it with?' It's so much fun."

The massive Building 9 at Johnson Space Center houses full-sized mock-ups of the individual ISS modules, and serves as a place not only for training but also troubleshooting during missions.

"We'll be in the middle of a shuttle mission and something doesn't work, so you'll have four OSOs sitting over there at 3 a.m. trying to figure out what to do," Martin explains.

She says flight controllers are great at the art of prioritizing properly, and asking the right questions to best determine how and when a problem onboard the ISS can be fixed.

"Things go wrong all the time — it's just a matter of the business," she says. "But during critical timelines, like during a shuttle mission, where you have to do things by a certain time or it can affect the next EVA [extravehicular activity], it's all about criticality, knowing how long we have to figure out a solution and what course of action we can take. Sometimes the answer is that we have to fly a new piece of hardware up on the next shuttle mission."

OSOs don't just address emergencies, though. They also research and write intricate procedures for operations such as installing new modules, and then train the astronauts on how to carry out those incredibly detailed lists of instructions. Every time a new module is launched to the ISS, the OSOs are involved in developing all the outfitting procedures to attach it and get it up and running.

Martin headed up the procedure development for the installation of the Cupola module, the "bay window on the Earth" that was launched on STS-130 in February 2010.

> ## "It's like Apollo 13. We sit down and say, 'OK, this didn't work. It wasn't designed right — so what do we have onboard that we can fix it with?' It's so much fun."

"I worked for months on that module," she says. "And to see those window shutters finally open was really amazing."

Martin began with a list of tasks necessary to get the Cupola up and running and the software working, but she had to determine the proper order in which those steps had to occur and then train the crew on exactly how to carry them out while on orbit. It's a high-stakes DIY tutorial.

"Imagine someone giving you a list of what has to be done to fix a car, and you're sitting there with this list trying to do it," she says. "That shows you how important the procedures have to be. They have to have good pictures, clear instructions."

Rachel Hobson is a hand embroidery-obsessed crafter and writer with a passion for outer space and all things geektastic. She is a staff writer for CRAFT (craftzine.com) and has her own blog, averagejanecrafter.com.

High-Resolution Spectrograph

Lab-worthy spectrum analysis for cheap.

BY SIMON QUELLEN FIELD

A

Nearly 200 years ago, Joseph von Fraunhofer built the first spectroscope and saw dark lines in the spectrum of the sun. This led him to discover that you can determine the chemical elements in things by analyzing the light they give off. Each element has its own "signature" of lines in the light spectrum. The lines correspond to the characteristic wavelengths that electrons absorb and emit as they jump between lower- and higher-energy orbits around the atom's nucleus.

The first spectroscopes used glass prisms to split light into colors, but Fraunhofer found that an array of closely spaced wires had the same effect. Today we call these arrays of tiny slits *diffraction gratings* (see sidebar on page 61).

After these discoveries, spectrographs (the instruments used to record spectra) became a standard tool for analyzing the chemistry of almost anything, ranging from microscopic lab samples to faraway galaxies.

The primary element in both spectroscopes and spectrographs is a narrow slit oriented perpendicular to the direction in which the grating or prism spreads the light. As with a pinhole camera, the small aperture images the light source sharply along the spectrum's axis, which keeps the spread

of wavelengths distinct. Each image of the slit, in a slightly different color, is arrayed across the field of view in a wide spectrum image. If any wavelength is brighter or dimmer than the rest, it shows up, respectively, as a bright or dark line in the spectrum.

Although spectroscopes have always been easy to make, a homebrew recording spectrograph presented more of a challenge. Building your own spectrograph meant using microcontrollers and stepper motors to move diffraction gratings past a light sensor — many were planned, but few were actually built.

Today, digital cameras and online tools can turn a simple spectroscope into a laboratory-quality, high-resolution spectrograph. All it takes is a few plumbing parts and other inexpensive materials and less than an hour at your kitchen table.

Photography by Ed Troxell

MATERIALS

ABS plastic pipe, 2" diameter, 15" long
Most hardware stores will cut ABS pipe to length.
ABS angled pipe coupling, 2" diameter 22½° is ideal, but 45° will do in a pinch.
Rubber cap for 2" pipe
Hose clamp for 2" pipe
Holographic diffraction grating film,
1,000 line-per-millimeter enough to cut a 2" diameter circle; item #DIFFRACTION at Scitoys Catalog (scitoyscatalog.com), $3
Black construction paper, 8½"×11"
Scrap cardboard some thin (like a cereal box or business card) and some thicker (like a shoebox or the back of a writing pad)

TOOLS

Digital camera
Computer with internet connection
X-Acto knife or razor blade
Ruler
Drawing compass and pencil
Glue Elmer's white glue, Duco cement, or similar

Build a High-Resolution Spectrograph
Time: Less Than 1 Hour
Complexity: Easy

1. Make a spectroscope.

Use a ruler and X-Acto knife to cut a straight slit about 1¾" long through the center of the rubber cap. Cut two ½" squares of the thin cardboard and tuck them into opposite ends of the slit to hold it open. The slit is now as wide as the cardboard.

Secure the rubber cap on one end of the 2" tube with the hose clamp. Rotate the hose clamp so that its screw is perpendicular to the slit (Figure B).

Loosely roll up the black construction paper lengthwise and slide it all the way into the tube so it fits neatly against the inside. This will serve to eliminate reflections. Use a compass to draw a ring on the thicker cardboard with an outside diameter of 2" and an inside diameter of about 1⅝", and cut out the ring using the X-Acto knife (Figure C). The ring should fit inside the tube snugly but without bending.

Spread a very thin layer of glue onto one side of the cardboard ring and affix it to the diffraction grating (Figure D). After the glue dries, carefully cut the diffraction grating around the outside edge of the ring using the X-Acto knife.

Photography by Simon Quellen Field

Insert the diffraction grating ring into one end of the angled coupling so that the ring's edge is flush with the interior wall, and press the uncapped end of the tube in until the ring is held in place. That's it — the tube is now a spectroscope!

To look at spectra, point the slit end up to a light source, look into the angle fitting, and rotate the rubber cap (or angle fitting) until the spectrum you see is a neat rectangle instead of a parallelogram or a thin line.

With the hose clamp's screw perpendicular to the slit, the clamp acts as a stand, keeping the slit vertical while preventing the tube from rolling.

You can experiment with the cardboard spacer thickness. A wider slit will make a brighter image with broader lines, giving less resolution. Using a longer or shorter tube will make the slit look narrower or wider, respectively.

2. Turn it into a high-resolution spectrograph.

To turn the spectroscope into something that records and analyzes the light it diffracts, we need a digital camera and the internet. Aim the camera into the spectroscope so it captures any spectrum you want to analyze, and set the zoom so it fills the viewfinder as much as possible to get the maximum resolution (Figure E).

To capture the spectra of stars, you don't need

the slit. The star is a point source of light, so it acts as its own slit. Just put some grating over the telescope eyepiece, and photograph the rainbow streaks on either side of the star.

Take several pictures of the spectrum, and if your camera lets you set exposure times and apertures manually, use a range of these settings. A properly exposed image will be dark, especially if you're capturing an emission spectrum, like from a fluorescent lamp.

Avoid the temptation to take a colorful rainbow picture, in which all the lines will be smeared out (Figure F). With the fluorescent light, for example, you want just 4 or maybe 6 of the mercury lines to stand out visibly (Figure G).

Download the photos onto your computer and crop them so that just the spectrum itself is visible. To analyze a spectrum image, upload it using the "Simple spectrum analyzer" form at my website at makezine.com/go/spectrograph (Figure H).

If you upload your spectrum photo without clicking on either checkbox, you'll get a simple spectrum plot. With my fluorescent lamp spectrum, for example, the graph shows 4 peaks. These peaks correspond to the brightest lines in the spectrum of mercury, which exists in vapor form inside the tube (Figure I).

You can use a spectrum source that's known, like our fluorescent light, as a high-precision ruler for

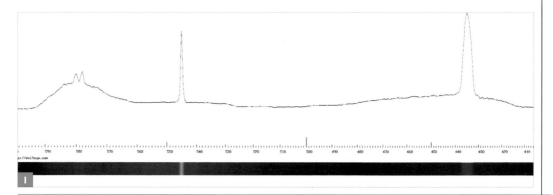

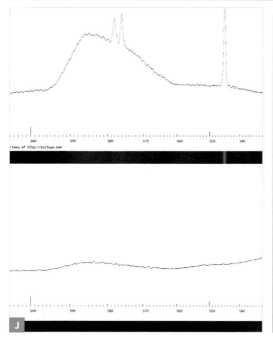

HOW DIFFRACTION GRATINGS WORK

Our diffraction grating is a transparent sheet of plastic with dark lines on it spaced 1/500 of a millimeter (2µm) apart.

Light waves pass through the spaces between the lines, called slits, and interfere with waves going through adjacent slits to produce bands of light or dark where the interference is constructive or destructive.

Light bands form where the wavecrests from adjacent slits are both an integer number of wavelengths away from the diffraction grating. Since this depends on the wavelength, it causes different colors to form bright bands in different places, separating them into the rainbow we call the spectrum.

analyzing other spectra, such as through a sample of transparent material or liquid.

To do this, you position the reference light to shine through the top half of the slit, and shine a broad-spectrum light source through your sample, covering the bottom half. Use a bit of aluminum foil to divide the slit into two parts, separating the calibration light from the sample light.

As an example, I used the fluorescent tube as a reference for testing some green clear plastic lit from behind with a white LED.

Upload the double spectrum image to the spectrum analyzer with the first checkbox checked, and you'll get a plot of both against each other. This analysis results in 2 graphs: the calibration on top, like the graph we just looked at, and the sample

spectrum on the bottom (Figure J).

The plots have a nanometer scale at the bottom, which we can verify against the 4 brightest lines in mercury's emission spectrum, as listed in the *Handbook of Chemistry and Physics* (CRC Press, 2009): 435.8nm, 546.1nm, 577nm, and 579.1nm.

Since both sources pass through the same slit and the same camera, calibrating this way guarantees that the frequencies in each spectrum match up exactly, and therefore that the readings are accurate.

Simon Quellen Field (sfield@scitoys.com) is president and CEO of Kinetic MicroScience (scitoys.com), where he designs scientific toys. He's the author of several books on science and computing.

Rocket Men

Mavericks of the private space industry.
BY CHARLES PLATT

FINAL FRONTIERSMEN: (Left to right) Dave Masten, John Carmack, Tim Pickens, Tim Bendel, Paul J. Breed, and Jeff Greason.

It's the best of times and the worst of times for space enthusiasts.

Following the imminent retirement of the space shuttle, the United States will lose its manned orbital capability. President Obama cancelled the Constellation program, which was supposed to take us back to the moon. An astonishing $9 billion was spent on Constellation, and canceling it will cost another $2.5 billion.

Yet amid this gloom and doom, a rapidly shifting mosaic of startups is maturing, hoping to get into orbit with a tiny fraction of what NASA has spent. Some quick case histories:

» Former Cisco software engineer **Dave Masten** started Masten Space Systems in the Bay Area in 2004 and later moved it to a 60-year-old wooden hangar at Mojave Air and Space Port in the California desert. In 2009, his tiny team of enthusiasts demonstrated an unmanned vehicle that launched vertically, moved laterally, and landed vertically. They took home more than $1 million as winners of the X Prize Lunar Lander Challenge sponsored by NASA and Northrop Grumman.

» Armadillo Aerospace was founded in 2000 by **John Carmack**, technical director of Id Software, famous for its *Doom* and *Quake* computer games. He hooked up with members of the Dallas Area Rocket Society and pursues space as a part-time occupation. Armadillo won the $500,000 second prize in the 2009 Lunar Lander Challenge.

» Maverick Alabama rocket enthusiast **Tim Pickens** founded Orion Propulsion in 2004. In 2009 he sold the company to Dynetics, a science and technology company employing more than 1,300 people and developing microsatellite capability. He's now their commercial space advisor and chief propulsion engineer.

» Six years ago, former Lockheed Martin engineer **Tim Bendel** sold his house outside Denver, bought a decommissioned Atlas missile silo in Wyoming, and moved into it with his wife and business partners. Their company Frontier Astronautics offers rocket engines, guidance systems, propellant, and a safe testing environment to other startups.

» Small businessman **Paul T. Breed** and his son, Paul A., formed their company Unreasonable Rocket in 2006, hoping to win the Lunar Lander Challenge. They didn't succeed, but built and flew three vertical-takeoff vehicles, built and fired nine types of rocket motors, and built an autopiloted helicopter to test their control software.

These startups face significant barriers, including scarce venture capital, tightening regulatory controls, environmental issues, and anxiety about safety. "This country is losing the ability to understand that risk-taking is wonderful," laments Shariar "Jack" Ghalam of Frontier Astronautics. "If you want to go into space, you're going to lose lives and hardware. If you're not willing to face that, just stay

home and do basket weaving."

Profiled here are a few of those who think that the risk-taking will be worth it.

The Spaceport Manager

Mojave Air and Space Port is a uniquely American resource: a funky, informal refuge for more than 40 aerospace ventures. For 35 to 50 cents per square foot, you can rent anything from a one-person office in which to write attitude control software, to a hangar big enough to build your own spaceplane.

The Voyager Restaurant, appended to the old control tower at Mojave, provides a grandstand view of the adjacent runway. Here you may run into the spaceport's general manager, **Stuart Witt**, who used to be a top-gun F-18 pilot back in the day. His mission is simple: "I offer people the freedom to experiment."

Suppose you come here wanting to test your own rocket motor. Witt doesn't do design evaluations. Once you've signed your contract, he just needs to know two things: what kind of fuel you'll be using (so he can put out the fire if the thing blows up) and how many people will be at the test stand (so he knows how many bodies to look for if people get killed). The rest is up to you.

"If you give people freedom and responsibility,

Fig. A: The Masten XA-0.1E, nicknamed Xoie, was flight-tested from this concrete pad (middle distance) near Mojave Air and Space Port (far distance) while team members monitored it from a control room in the industrial shipping container behind these concrete blocks. Xoie won the million-dollar Lunar Lander Challenge in November 2009.
Fig. B: The Voyager Restaurant at Mojave Air and Space Port, named after the aircraft designed by Burt Rutan that made a record-breaking nonstop flight around the world in 1986, copiloted by Jeana Yeager and Burt's brother, Dick. Burt's company, Scaled Composites, is based at Mojave.
Fig. C: Burt Rutan's SpaceShipOne was the first privately built, flown, and funded craft to reach space, winning the Ansari X Prize in 2004. This full-sized replica is on display in a small museum at Mojave Air and Space Port. Fig. D: Xcor Aerospace plans to offer suborbital tourist flights in its Lynx spaceplane, currently under development in this wooden hangar on the flightline at Mojave Air and Space Port.

they rise to it," Witt says. "If you want to be their nanny, they expect you to take care of them — and everybody loses."

Masten Space Systems tested its winning lander at Mojave, out near the edge of the airport in a parched area of desert scrub. A concrete slab about 20 feet square is scarred with burn marks at the center, showing where the vehicle took off and

Photography by Charles Platt

landed. Fifty feet or so behind the pad is a stack of massive concrete blocks, which shielded the launch crew during tests. An industrial shipping container housed the control room. Cheap? Absolutely. But it got the job done.

The Rocket Test Pilot

Dick Rutan received the Presidential Citizens Medal for copiloting the Model 76 Voyager around the world without stopping or refueling in 1986, an historic first. Long before that, he served for two decades in the Air Force and flew 325 combat missions over Vietnam. He has the Silver Star, five Distinguished Flying Crosses, 16 Air Medals, and the Purple Heart.

He's also flown a privately funded rocket-powered vehicle that has emerged from Mojave: the EZ-Rocket, a kit-built airplane designed by his brother, Burt, with an Xcor rocket engine (see The Businessman, page 66) shoehorned into the rear. "I was really intrigued by that," he recalls. "I flew it about 14 times. I never got paid for that, to my wife's chagrin." He chuckles.

Rutan is a quintessential plainspoken individualist for whom failure is abhorrent. "If you come in second," he says, "you're banished to the lake of

Fig. E: The tiny town of Mojave, Calif., houses its Chamber of Commerce in this converted railroad car. Fig. F: The EZ-Rocket spaceplane. Xcor shoehorned its rocket engine into a refurbished Long-EZ airplane for flight testing. The plane was designed by Burt Rutan. His brother, Dick, test-flew the conversion. Fig. G: In Xcor's wooden hangar, a mockup of the cockpit of its upcoming Lynx suborbital vehicle has been fashioned from fiberglass. Fig. H: The test stand for an Xcor rocket motor.

obscurity forever."

And he is utterly committed to space exploration. "In America, the Apollo program is the greatest thing we ever did. A young president wrote the check and then got the f--- out of the way, and the people who were involved in that did an incredible thing. We went to the f------ moon! Our leadership today, besides being naive and incompetent, squandered the capability of a great nation, because they think that food stamps are more important than going to Mars — the most naive, idiotic policy that I can even imagine. The viability of a country depends on its technology."

SpaceShipOne, designed by Burt Rutan and built by Scaled Composites, won the $10 million Ansari X Prize for the first privately funded human space-

Fig. I: Dick Rutan works on his Berkut, a kit-built aircraft descended from the Long-EZ design developed by his brother, Burt. Fig. J: A rocket motor stands on a workbench in a Masten Space Systems hangar at the Mojave Air and Space Port. The motor was turned from a block of aluminum at a local metal-fabrication shop. It is double-walled, enabling circulation of coolant to prevent the aluminum from melting. Fig. K: The XA-0.1-B (generally known as Xombie) won second place in Level 1 of the NASA-funded Northrop Grumman Lunar Lander X Prize Challenge in 2009. Xombie is still being used for flight testing new systems.

a barrier — do not ever comply with it.

"Look at it," Rutan says, "as a target of opportunity for greatness."

The Motor Man

Tim Pickens sounds like a NASCAR driver, which isn't such a far-fetched comparison, since he owns a home-built rocket-powered pickup truck. He also built a rocket-powered bicycle, using a hybrid booster fueled with asphalt and nitrous oxide.

"I tried to run a company," he says, referring to his Orion Propulsion startup, which earned the enviable distinction of being one of the few small, private space ventures that has actually made money. "But I always want to do this fun stuff."

As a rocket-obsessed amateur searching for fellow travelers, he went to a National Space Society meeting in 1992 and was shocked to find that nobody there had actually launched anything.

"I'd been flying my 11-foot, 55-pound steam rocket," he recalls. "It would get 500 pound-seconds, and that baby would get up there fast. I had 22 pounds of water in there, and a fire extinguisher as a tank. I heated it to about 400psi. I used to test the steam motor in my driveway."

Hoping to put together some "smart engineers" with higher ambitions, he moved to 120 acres of

flight. It used a unique pair of swiveling booms to raise the tail and create drag during reentry. The original vehicle is now in the Smithsonian — which is where Dick Rutan thinks it should stay.

"They had some really close calls," he says. "We had fires and problems, and it was damned dangerous sometimes. But nothing worthwhile was ever accomplished without risk or daring."

He continues: "My message to young people is, your responsibility is to arm yourself with the basis of education to give yourself the tools to do something that nobody here can even imagine. But you need to know the physics of how things operate. And then, if you look around and see a limitation or

"We can't ride on the back of Apollo forever. We got to do something."

vacant land owned by his father. "I got some great videos," he drawls. "Like, you know, unintended launches, and, uh, dynamic disassembly."

He pulls out his iPhone to show pictures of his big machine shop, which he refers to as his "man cave," and harbors no illusions about his pathology.

"I'm a fanatic. Once I get interested in something, I bore in, man. See, I wondered — why is this stuff so hard to build? I made it simple. No hand-laid-up composites and hand-welded components and cryogenics. I avoided sliding surfaces of components with high thermal coefficients. I had to ask, why are we making things too complex?"

He started Orion with a savvy business model. Rather than try to build a whole rocket, he figured he would specialize in engines, which he knew about, and sell them to aerospace companies. "Like selling shovels to miners," as he puts it. "When you become a miner, there's a lot more risk, as you may not strike gold. If you sell shovels, you're always going to make a little money."

He had a profitable five years, but when he got a good offer from Dynetics to take over his company and almost all its employees, he had no hesitation. He's in the corporate culture now, but still tackles fun projects at home. His daughter, Sarah, 18, test-piloted his most recent invention, a water-powered rocket belt attached via a 4-inch water hose to a Jet Ski. He bought almost all the parts from Home Depot.

Behind the stunts, though, he's deadly serious about wanting to go into space. "We got all that heritage, man," he says. "But we can't ride on the back of Apollo forever. We got to do something."

The Businessman

Jeff Greason is calm, amiable, and endlessly diplomatic. From 1988 through 1997 he worked for Intel. Still, as he puts it, "I had gotten passionately interested in commercial space transportation. And I couldn't see spending the rest of my life worrying about how to get Intel from 85 to 90 percent market share."

He took a substantial financial hit when he quit his job and founded Xcor Aerospace. It now has around 20 employees and has developed something that was thought to be impossible:

a piston-based liquid oxygen pump. The technology lies at the heart of Xcor's plans for a rocket plane named Lynx, to take paying customers on suborbital flights. Greason expects to sell tickets for less than half the $200,000 touted by Virgin Galactic, and summarizes the mission profile in three short sentences: "Fifteen seconds from brake release to rotation. Fifty seconds to supersonic. Three minutes of weightlessness."

Meanwhile, Xcor has partnered with United Launch Alliance to develop engine technologies for an upper stage for satellite launches.

Why is Greason so confident he can compete with established aerospace contractors? "A whole generation of aerospace engineers have no memory of what a competitive industry is like," he says. "Under a government contract, typically you get paid for cost plus a small percentage. Therefore, if you find a way to make the product cheaper, you get paid less!"

Since his competitors spend almost unlimited sums to develop vehicles that are used once and thrown away, he figures that "any reasonably economic reusable vehicle would change the basis of space transportation profoundly — if it just works."

Already he has demonstrated an aircraft with an Xcor rocket engine in it. This vehicle is capable of rapid turnarounds, like a regular jet aircraft. Scale models of the Lynx have been wind-tunnel tested, and the final design will carry one pilot and one paying passenger.

In a typically low-rent hangar at Mojave Airport, a full-scale mockup of the Lynx cockpit has been created, and fuel tanks are being made from layers of glass fiber, while aluminum components are being turned out in a machine shop next door.

"I'm confident that there's a robust market where the price for a flight is in the high tens of thousands of dollars," Greason says. "And we're the only company that doesn't have one rich guy paying for everything. We have individual investors. We're not a hobby for someone, we're actually a business."

His big technical advantage is proven reusability. "In a typical rocket engine, the inside wall gets hot and the outside is cool. You get thermocyclic fatigue, and the lifetime of those engines tends to

The Rocket Truck

be quite short. So we devised a series of design changes that accommodate thermal expansions and contractions. Since we got the design the way we wanted it, we've never worn one out."

Thus far, Xcor has received relatively little public attention. The company made a tactical decision not to go all-out to win the $10 million X Prize for private manned spaceflight that was claimed by Rutan's SpaceShipOne. "We decided we could do it right, or do it soon," he says, "but not both."

Some people have compared the X Prize to the Apollo program, describing both as "stunts" — one-goal endeavors with no commercial followup. Greason is too diplomatic to make such a comparison, but he's very clear about his goals.

"I am not interested in stunts," Greason says, quite firmly. "I'm interested in markets."

Special thanks to Robin Snelson for help in researching this article.

Fig. M: Gerry Mulryan, master mechanic at Xcor Aerospace, stands beside one of the tools of his trade. Fig. N: At Masten Space Systems, a rocket motor is mounted on a trailer, ready to be towed out to one of the test pads near the perimeter of the Mojave Air and Space Port. Fig. O: An older family photo of Tim Pickens and his daughter, Sarah (now 18), with rocket-powered bicycles that Tim designed and built. (Hers was powered by safe compressed carbon dioxide.) Fig. P: A PowerPoint slide from Pickens, showing his rocket-powered pickup truck.

➕ More on private space flight and manufacturing:

Rocket Sellers blog: rocketsellers.wordpress.com
Space Studies Institute: ssi.org
Scaled Composites: scaled.com
Masten Space Systems: masten-space.com
Frontier Astronautics: frontierastronautics.com
Paul Breed: unreasonablerocket.blogspot.com
Bigelow Aerospace: bigelowaerospace.com
Armadillo Aerospace: armadilloaerospace.com
Dick Rutan: dickrutan.com
Tim Pickens: realrocketman.tripod.com/uncletim.htm

Charles Platt is a contributing editor to MAKE.

Map the Chemical Composition of the Moon

Boost the saturation of digital photos to reveal lunar geology.

BY MICHAEL A. COVINGTON

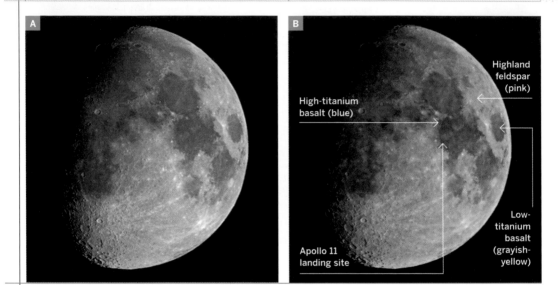

A

B

Highland feldspar (pink)

High-titanium basalt (blue)

Low-titanium basalt (grayish-yellow)

Apollo 11 landing site

The moon is more colorful than most of us realize. Although the colors are barely visible to the naked eye, this simple digital photography technique makes them obvious. Not only are they pretty, they reveal the chemistry of the lunar surface.

Take Your Picture

You can photograph the moon using a telephoto lens or a telescope adapter, or by simply holding your camera up against the eyepiece of a telescope (5 to 50 power). Either the camera or the telescope needs a tripod. Holding the camera by hand seems kludgy, but it works well because the telescope's mass reduces the camera's vibration.

If you're holding the camera, zoom to the middle of its optical zoom range, disable its autofocus, and adjust the telescope's focus while looking at the camera's display. If the camera is physically attached to the long lens or telescope, use the time-delay shutter release so that everything can settle down after you take your hands off it. Non-SLR digital cameras usually don't produce much

vibration, but the shutter and mirror in a DSLR do. With Canon DSLRs, using Live View mode ("silent shooting") eliminates this vibration by starting the exposure electronically rather than by moving the shutter.

Using manual exposure, set the aperture wide open (f/2.8) and try 1/200 second at ISO 200. But experiment — try for a picture that's a bit under-exposed, so that the brightest areas don't become colorless white. And use automatic white balance, so that the image won't have an overall color cast.

Above all, select a TIFF or raw (proprietary) file format. A 16-bit TIFF is best. Don't select JPEG compression, which discards exactly the subtle colors we're trying to bring out. Converting a JPEG-captured image back to TIFF doesn't undo the damage.

Photography by Michael A. Covington

You can also scan a color film image, even one you took many years ago. As you process the picture, look out for "crossover" — color casts that are different in brighter than in darker areas. Crossover was the main bugaboo of color film and one reason that digital photography displaced it so quickly.

Do the Magic

Bring the image into Photoshop or another image-processing application, then simply increase the color saturation. In Photoshop, this adjustment is accessed via Ctrl-U on a PC and Command-U on a Mac.

Look at the results as you turn up the saturation. Don't go too far — if the image becomes grainy or gaudy, you've gone beyond the colors actually recorded from nature and are now looking at camera artifacts. It may work best to make the adjustment in 2 or 3 steps.

Finally, you may want to do some sharpening or unsharp masking to bring out lunar details. After processing, you can save your picture as a JPEG because the low-level colors are no longer hidden.

What You'll See

There are three basic kinds of moon rocks: highland material (feldspar), low-titanium basalt, and high-titanium basalt. These come out as pinkish, grayish-yellow, and blue, respectively. (The powder, or regolith, on the lunar surface is mostly similar to the rocks beneath, although it also contains material from micrometeorites.)

Note especially the blue color of the Mare Tranquillitatis (aka Sea of Tranquility), contrasting with the other lowlands. Apollo 11 landed in an especially blue patch here because it is flat terrain with few boulders to run into.

Greenish tints can indicate places with higher iron concentration, although the edge of the sunlit region can also appear green because that's how the rocks look when lit from the side.

Other unusual tints have been reported in Sinus Iridium (aka Bay of Rainbows) and around the crater Aristarchus, which has a reputation for odd behavior — it reflects sunlight so brightly that it's been mistaken for an erupting volcano.

By day, Michael A. Covington is a senior research scientist at the University of Georgia. By night, he photographs the sky and writes books about it, such as *Digital SLR Astrophotography* (dslrbook.com).

Photograph by Simon Quellen Field

DIY Ion Engine

BY KEITH HAMMOND

To get to Mars, we'll use the same technology as Darth Vader's TIE fighter — the ion propulsion engine. It uses electricity to produce a plasma of charged ions exiting 10 times faster than chemical rocket exhaust.

The most powerful ion engine today is Ad Astra Rocket Company's 200-kilowatt VASIMR plasma engine, which uses radio frequencies to superheat argon ions to 1,000,000°F. Ad Astra's founder, former NASA astronaut Dr. Franklin Chang Diaz, estimates a nine-month one-way trip to Mars could be cut to 39 days using 200-megawatt plasma engines.

Until then, Simon Quellen Field shows how to make your own tiny ion motor using just a high voltage source and two paper clips, at his website Science Toys You Can Make With Your Kids (makezine. com/go/ionmotor). The journey of 100 million miles begins with a single ion.

TINY ION: Replicate the basics of an ion propulsion engine using a piece of wire, the cap from a D battery, and a power source.

Space Rock Hounds

A Q&A with the Meteorite Men.

BY RACHEL HOBSON

Geoff Notkin and Steve Arnold hunt for visitors from outer space, but they aren't looking for little green men. Their treasure is of the geologic sort: meteorites that strike Earth, carrying clues to astronomical and planetary development in the universe. The pair share their adventures in the Science Channel's TV series *Meteorite Men*. We checked in with them to get their tips and insights for DIY meteorite hunters.

Geoff Notkin of *Meteorite Men* in his rover.

Q: What's the simplest form of meteorite hunting that the DIY enthusiast can tackle?

A: Meteorites fall randomly over the entire surface of the Earth, so theoretically you could search for them anywhere. The vast majority of meteorites contain a large amount of iron, which rusts in humid environments, eventually causing the meteorite to decompose. Head out to a barren area that's devoid of vegetation, with few indigenous rocks, and see if you can spot anything unusual.

Q: What kind of research should you do before setting out on a meteorite hunt?

A: Many types of common Earth rocks are frequently mistaken for meteorites. Become familiar with what meteorites look like, and also practice a few simple field tests that can help in identifying suspected space rocks. Our Guide to Meteorite Identification [aerolite.org/found-a-meteorite.htm] is a good place to start.

Q: What kind of meteorite-hunting tools can you make at home, instead of having to purchase them?

A: We end up designing and building much of the equipment that we use in the field. The simplest tool, and one of the most effective, is a magnet cane. It's basically a walking stick with a powerful magnet affixed to one end. When you're hiking 10 or 15 miles a day, keeping your eyes peeled for unusual-looking rocks, it's a time saver and an energy saver if you don't have to bend down and pick up every one of them.

➕ **For an extended interview with the Meteorite Men, visit: makezine.com/24/space**

Lunar Lander Simulator

BY JOHN BAICHTAL

INSPIRED BY ATARI'S 1979 *Lunar Lander* game, British engineer Iain Sharp created a lunar landing arcade machine to honor the 40th anniversary of the historic Apollo 11 mission.

Players work a steering wheel and rocket controls to safely bring the simulated Eagle lander to the moon's surface while keeping an eye on the speed and fuel indicators.

But this isn't a video game — Sharp's lander dangles from fishing line manipulated by salvaged inkjet servos. Controlled by Atmel AVR microcontrollers coordinated by an Arduino, the simulator uses infrared sensors to determine the lander's location. For a retro Space Age look, it sports vintage-looking instrumentation and nixie-tube readouts.

Iain Sharp's 2009 *Lunar Lander* console game uses physical components instead of video.

Photography by Suzanne Morrison ©Aerolite Meteorites (above); Iain Sharp (below)

Zillionaires in Space BY LAURA COCHRANE

Photograph by Adam Goss (planetarium)

It's not absolutely necessary to be rich if you want to explore space, but it sure doesn't hurt. Here are some financially well-endowed individuals who don't mind spending part of their fortunes getting into orbit.

Jeff Bezos
Amazon.com's billion-aire founder has sold off $640 million of stock this year while bankroll-ing his Seattle-based aerospace company, Blue Origin. They're developing escape systems for NASA rockets and planning their own suborbital space tourist flights by 2012 from their West Texas launch site.

Sir Richard Branson
The British billionaire behind the Virgin mega-brand founded Virgin Galactic in 2004, and is selling $200,000 tickets for suborbital passenger flights, by 2012 or so, from their spaceport in New Mexico.

Branson sees space travel as a way of speed-ing up transcontinental travel: "We'll be going city to city by just shoot-ing people out to space and straight back down again."

Elon Musk
The South African founder of PayPal and Tesla Motors launched California rocket maker Space Exploration Technologies (SpaceX) in 2002.

Despite high-profile setbacks (some of *Star Trek* actor James "Scotty" Doohan's ashes met their final frontier in the Pacific), the company proposes to transport payload and crew to the International Space Station (ISS), and to reach Mars with nuclear-thermal rockets.

Peter Diamandis
The X Prize founder is also a director of Virginia-based Space Adventures, the com-pany that's flown seven well-heeled tourists to the ISS, including video game mogul Richard Garriott and telecom millionaire Anousheh

Ansari, for prices of $20 million to $35 million.

They're now booking suborbital flights too, as well as $100 million seats for a circumnavi-gation of the moon.

John Carmack
The founder and top developer of Id Software (*Quake*, *Doom*, and the space-themed *Commander Keen*) launched Texas-based Armadillo Aerospace in 2000.

Now they're partnering with Diamandis' Space Adventures to develop suborbital space tourist flights for an estimated $102,000 a seat.

Sergey Brin
While Google sponsors the $30 million Lunar X Prize, its billionaire co-founder is person-ally headed for space. In 2008 he invested $5 million in Space Adventures, as a down payment on an estimated $35 million ride to the ISS aboard a Soyuz launched in his native Russia, as early as 2012.

DIY Inflatable Home Planetarium
BY GOLI MOHAMMADI

Most home planetarium enthusiasts use a geodesic dome design, but Adam Goss says geodesics fall short because of their flat surfaces. The gore dome design (made from curved sections, or gores) makes a

truer hemisphere, and is used in many planetariums worldwide.

In 2009 Goss posted a tutorial for making a 5-meter inflatable gore dome. His site features free construction manuals for domes ranging from 3 to 5 meters, along with the GoreDome Calculator software that he developed for designing your own optimal planetarium dome. diyplanetarium.com

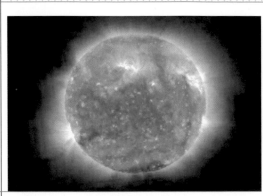

Five Cool Participatory Space Projects

Find new galaxies and design spacecraft. BY ARIEL WALDMAN

From launching your own satellite to analyzing aerogel, here are five ways you can make scientific discoveries and actively contribute to space exploration.

Galaxy Zoo
Classify and potentially discover new galaxies with Galaxy Zoo, a data set of thousands of galaxies imaged by the Hubble Space Telescope. With a simple interface, a community of over 250,000 collaborators, and easy tutorials, this is a great way for kids and adults to participate in space exploration. galaxyzoo.org

Team Frednet
Google Lunar X Prize is a $30 million competition to build and send rovers, robots, and spacecraft to the moon.
Team Frednet is an open source team competing in the challenge, and anyone can join as a contributing team member leading up to the rocket launch. frednet.com

« Solar Stormwatch
Coronal mass ejections on a collision course with Earth send dangerous radiation to astronauts and can knock out communication networks and power lines.
Learn how to spot these solar explosions using near-real-time data from NASA spacecraft. solarstormwatch.com

Collaborative Space Travel and Research Team
CSTART organizes and finances the efforts of space enthusiasts around the world who use collaborative design, volunteer labor, innovative low-cost technology, and open data sharing to further the cause of manned and unmanned space exploration and research.
CSTART creates open source plans for projects ranging from sounding rockets (the OHKLA project) to manned moon landings (the CLLARE project). They also write open source software to help plan space missions and to run onboard spacecraft. cstart.org

Citizen Sky
Epsilon Aurigae is a supergiant star located 2,000 light-years from Earth that mysteriously gets eclipsed every 27.1 years by an equally large, unknown, dark object. The event has baffled scientists for 175 years, but through the Citizen Sky project, you'll make observations and analyses that could decipher this scientific puzzle! citizensky.org

➕ Want to check out more cool participatory space projects? Explore the directory at Spacehack (spacehack.org).

Saturday Morning Science in Space

BY GOLI MOHAMMADI

WHEN CHEMICAL ENGINEER and NASA astronaut Don Pettit spent six months aboard the International Space Station in 2002–2003, he used his off time productively. Though most of the week was busy with research and maintenance, Saturday mornings were allotted as personal time, and Pettit used the micro-g environment to conduct and document a series of fun experiments for kicks, which he called Saturday Morning Science.

The footage is fascinating, informative, and available for free on YouTube. Using what was available on the ISS (like tea leaves, sodium chloride, and honey), Pettit demonstrates Marangoni convection, nucleate boiling, and gyroscopic platform stability, to name a few. The section on making water spheres is mesmerizing. Pettit even shows that motor-skilled repair work is possible in micro-g by repairing his broken watch. Tune in at makezine.com/go/pettit.

Photograph courtesy of SECCHI beacon (above)

Cash Prizes for Space Scientists

Solve a challenge, become the first space millionaire on your block! BY JOHN BAICHTAL

The Ansari X Prize for Suborbital Spaceflight (space.xprize.org/ansari-x-prize) paid $10 million to Scaled Composites LLC when its SpaceShipOne exceeded 100km in altitude. NASA realized entrepreneurs could develop technology far more efficiently, and the agency now offers tantalizing prizes to spur development of space tech.

NASA's Space Elevator Games (spaceelevator games.org) provides prizes to stimulate the development of a space elevator — a line stretched from the ground to orbit, with a climbing elevator car that delivers cargo to space cheaper than a rocket. In 2009 the agency awarded $900,000 to Seattle startup LaserMotive, who created an efficient way to beam electricity to a climber.

NASA's Regolith Excavation Challenge (regolith. csewl.org) awarded $750,000 to three teams that designed robots to dig lunar soil. The winning robot, created by college student Paul Ventimiglia, dug up 439kg of simulated regolith within 30 minutes.

Also in 2009, Peter Homer won NASA's Astronaut Glove Challenge (astronaut-glove.us). Homer took home $250,000 in prize money, on top of $200,000 he earned in the 2007 competition. He has started his own spacesuit glove company.

Interested in getting in on the action? Here are some contests:

STUDENTS
NASA Great Moonbuggy Race
moonbuggy.msfc.nasa.gov
High schoolers and collegians design collapsible, human-powered moon buggies.
Purse: Bragging rights

MoonBots moonbots.org
Mixed teams of kids and adults create simulated lunar robots out of Lego Mindstorms NXT sets.
Purse: A free trip to Lego HQ in Billund, Denmark

NASA Space Settlement Contest
settlement.arc.nasa.gov/contest
Middle and high schoolers design orbital habitations in this annual contest. Teachers may download instruction packs to integrate the science of space stations into their lesson plans.
Purse: $3,000

NASA Lunabotics Mining Competition
nasa.gov/lunabotics
Challenges college students to build a lunar excavator to dig up 10kg of regolith within 15 minutes.
Purse: $500–$5,000 per category

PROFESSIONALS
Google Lunar X Prize googlelunarxprize.org
Awarded to the first team to land a robot on the moon, travel 500m, and transmit back to Earth.
Purse: $30 million

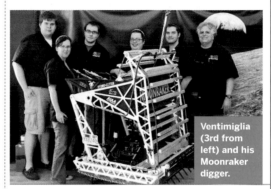
Ventimiglia (3rd from left) and his Moonraker digger.

Power Beaming Challenge
spaceelevatorgames.org
Compete to beam power to robots that climb up a cable, a critical development needed to someday assemble a space elevator.
Purse: $2 million

Strong Tether Challenge
spaceward.org/elevator2010-ts
Challenge to develop a lightweight cable strong enough to stretch from orbit to Earth.
Purse: $2 million

Sample Return Robot Challenge
centennialchallenges.nasa.gov
Rovers must autonomously locate and collect specific geologic samples.
Purse: $1.5 million

🔲 **Learn about other challenges at makezine.com/24/space**

Photograph by Jamie Foster/CSA

Space Science Gadgets You Can Make for NASA

BY MATTHEW F. REYES

NASA faces uncertainty not only about its mission, but about how to pay for it — the agency gets less than 0.6% of the federal budget. We must find ways to cut the cost of working in space.

Today the maker community is changing the economics of how to do-it-yourself in outer space, by hacking Android smartphones into tools that NASA could use for science discovery.

In December 2009, some Googlers started cellbots.com to create smartphone-controlled rovers. Two months later hackers at Noisebridge in San Francisco launched the first G1 Android phone aboard a high-altitude balloon.

In July 2010, volunteers at NASA Ames Research Center and Google launched a Nexus One aboard a suborbital fiberglass kit rocket from the site of Burning Man (see below). Two weeks later the education nonprofit Quest For Stars in San Diego launched a Motorola Droid aboard a balloon to more than 100,000 feet!

The joint cadre of volunteers at NASA Ames and Google are now hacking Android smartphone components to control robotic rovers, aerial vehicles, and small satellites, and to collect science data from them. Within a limited budget, we're relying on students and the maker community to create and program this new class of universal, open source participatory exploration platforms.

We need immediate help to develop wi-fi, Bluetooth, or USB interfaces that can connect scientific data collection payloads to devices running the Android OS. From there, there's boundless potential for makers to create previously unthinkable gadgets to support NASA's mission, including:

» "Mini-Hubble" space telescopes that can send space images to amateur astronomers

» Ruggedized "cellbots" that can explore extreme environments on Earth and on near-Earth asteroids

» Remote-sensing, environmental-sampling aerial vehicles, such as balloons and helicopters, that can help analyze climate change.

What can you imagine exploring with your smartphone? Let me know at motorbikematt@gmail.com.

Matthew F. Reyes is founder of Exploration Solutions, Inc., an education subcontractor at NASA Ames Research Center. twitter.com/motorbikematt

Androids at 28,000 Feet BY ADAM FLAHERTY

As part of a series of experiments to determine the viability of using off-the-shelf smartphone components in low Earth orbit (LEO) satellites, the "AndroidSat" project successfully launched a pair of Nexus One smartphones over Nevada's Black Rock Desert. Traveling to 28,000 feet aboard James Dougherty's Intimidator 5 rocket, their payload successfully recorded onboard sensor data and video of the launch.

Matthew Reyes (see above), Chris Boshuizen, Will Marshall, and the NASA Ames interns associated with the launch are championing the use of smartphone components to lower the cost of deploying a satellite, and they expect it to become even more affordable with each revision.

Photography by Ryan Hickman (above left); Andrew Gerrard (above right); Steve Jurvetson (below)

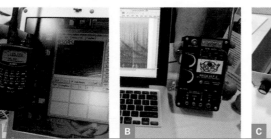

A

B

C

D

Live from Outer Space!

Tune in to heavenly sounds with amateur radio. BY DIANA ENG

Ham radio operators are no strangers to space communications. Astronauts use ham radio to communicate with hams on the ground, and operators use it to control satellites. If you've ever wanted to listen to the sounds of space — from Jupiter's natural "radio station" to text messages sent via the moon — you don't have to be a rocket scientist; you just need an amateur radio.

Here are a few radios you can build and/or operate to receive extraterrestrial radio signals.

Satellite Radio This Alinco DJ-G7 handheld radio (Figure A, about $300, alinco.com) and adjacent computer are receiving messages from the ARISSat-1, a homebrew satellite made by an international team of volunteers. Set to deploy from the International Space Station later this year, ARISSat-1 will broadcast a recorded message from school-children around the world, data from experiments onboard the satellite, and images from the satellite's camera, and will allow ham radio operators to communicate with each other via the satellite's repeater.

To receive radio signals from a satellite, you need a VHF/UHF FM radio. Attaching a yagi antenna helps (although it won't work with this particular beacon broadcasting to school-children) — and you can make your own. (See the yagi antenna project on page 48.)

Catch a Whistler
On this computer is a visualization of a natural phenomenon called a *whistler* (Figure B): a radio signal created by lightning, traveling through space, skipping from one side of Earth to the other. The Inspire VLF-3 radio kit ($120, theinspireproject. org) lets you listen to frequencies emitted by lightning, ranging from 0Hz to over 100kHz: nearby lightning signals sound like a crackling fire; lightning signals that have traveled 2,000km–3,000km sound like a skipping record; and whistlers sound faintly like bottle rockets. Add a $10 antenna and you're good to go.

Jupiter's Jukebox
The kit-built RJ 1.1 receiver (Figure C, $155, radiojove.gsfc.nasa.gov) can be used to make a radio telescope to observe the sun and Jupiter, both of which have magnetic fields that generate radio waves. The sun's transmissions are best heard between 18MHz and 200MHz, while Jupiter emits in the 18MHz to 40MHz frequency range.

Solar bursts usually last from 30 seconds to two minutes and sound like a light rain that becomes a downpour then lightens to a gentle pitter-patter. Jupiter's radio waves are influenced by both planetary conditions and its orbiting moons — long bursts sound like ocean waves; short bursts sound like wind flapping a flag.

In addition to the RJ 1.1 receiver, you'll need to build an antenna suited to receive radio signals from the sun or Jupiter (instructions are on the Radio Jove website).

Moonbounce
Here's a home radio station that KB2FCV (aka James Kern) is working on to communicate via the moon (Figure D).

In a "moonbounce," one operator sends a radio signal to the moon. The signal literally bounces off the moon and is received by a second operator, who can moonbounce a reply. The antenna is homemade, following the specifications at yu7ef.com.

🎵 **Hear radio signal audio clips from satellites, lightning, Jupiter, and beyond:** makezine.com/24/ space

Diana Eng is an amateur radio operator, fashion designer from *Project Runway*, and author of *Fashion Geek*.

Photography by Diana Eng (A–C) and James Kern (D)

Make: SPACE

SuitSat BY RACHEL HOBSON

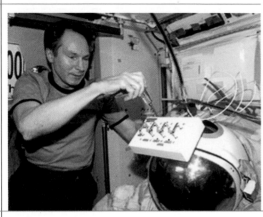

IN FEBRUARY 2006, crew members of the International Space Station hand-launched a disembodied spacesuit into Earth orbit during a spacewalk. While the sight of a human form floating away into space untethered was unsettling, the signals emanating from the space-suit satellite were meant to captivate the imaginations of people around the globe.

SuitSat-1 was a modified Russian Orlan spacesuit previously used for extravehicular activities. Engineers and ham radio operators outfitted the suit with equipment to broadcast prerecorded messages for downlink by listeners on Earth. But after just two orbits, operators began to lose contact with SuitSat-1, and it burned up on reentry in September 2006.

A planned SuitSat-2 was renamed ARISSat-1 when a second spacesuit couldn't be procured. Slated for launch in January 2011, ARISSat-1 will transmit radio messages and carry student experiments and solar panels that should help it run for at least six months. To learn more, check out arissat1.org.

Spaceport Sheboygan

BY KEITH HAMMOND

Need to catch a flight to space? Virginia and Florida both have licensed spaceports, and the West is peppered with them. But if you live in the Midwest, then Spaceport Sheboygan may be your ticket — a proposed spaceplane hub nestled on the shore of Lake Michigan.

The Wisconsin legislature created an aerospace authority in 2006 to promote a commercial spaceport for suborbital flights, currently projected by 2015. Why here? Restricted airspace left over from an old Army base provides a window to outer space, and the lake itself is handy in the event of a water landing. And when wealthy space tourists want to make a weekend of it, Sheboygan is a lovely town with great amenities: golf, boating, fishing, an annual air show, even surfing. (You can't say the same for Mojave, Calif., for example.)

First they'll need to extend a runway and apply for FAA approval, but Spaceport Sheboygan (spaceportsheboygan.org) is already home to amateur rocket launches (they've sent Lokis to 300,000

feet), Rockets for Schools, and the Great Lakes Aerospace Science and Education Center, featuring hands-on exhibits and summer space camps. Until the spaceplanes touch down, it's a great place to fly a shuttle simulator, check out astronaut gear that's flown in space, and watch the rockets streaking skyward while you're eating squeaky cheese curd.

➕ **All about spaceports:** hobbyspace.com/ SpacePorts

Photography by NASA/ISS program (above); William Friend (below)

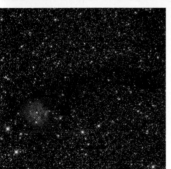

Backyard Astrophotography

Peter Shah brings distant galaxies down to Earth. BY RACHEL HOBSON

Photography by Peter Shah

When people see Peter Shah's breathtaking photographs of the Andromeda Galaxy and countless nebulae, they sometimes think they're from the Hubble Space Telescope. But Shah takes them from his backyard shed in Mid Wales in the U.K. He built it with a roll-off roof and a separate "warm" room (that isn't really very warm at all).

"The scope is a 200mm f/3.8, AG8 Newtonian Astrograph, on a Losmandy G-11 mount," Shah says. "I use a Sky-Watcher ED80 for guiding. The main imaging CCD is a Starlight Xpress SXVF-H16. The whole kit performs very well for a moderate imaging setup, but having nice dark skies does help too."

Shah doesn't pretend to be a whiz at math or astronomy, but says he learns as he goes along.

"The images that I produce are more art than science," he says. "By the time I've stretched the contrast, changed the levels, and enhanced the certain areas, the whole image becomes really a lot different to what it actually is. But I do think these sorts of images are very important to astronomy, as I truly believe it's the artists that inspire the young scientists."

📷 **More: astropix.co.uk**

1+2+3 Dizzy Robot
By Steve Hoefer

Dizzy Robots are cute pocket-sized pals that dance around until they fall over. Just about anyone can build one — it only has 3 parts and requires no special skills.

1. Prepare the metal body.

The metal body holds everything together and conducts power from the bottom of the battery up to the motor.

Trace the pattern shown at right, tape it to a thin piece of tin, and cut it out using tinsnips or heavy-duty scissors. Be careful, the edges and corners will be sharp!

Bend the square base of the body at a right angle, then bend the bottom pair of wings into a rough circle to hold the battery in place. Bend the top part of the body into a circle to hold the motor. Use a pen or pencil as a rough guide to help form the shape.

2. Prepare the motor.

If the motor came with a rubberized insulating cover, remove it. Use needlenose pliers to carefully bend one of the motor's contacts around and under the motor. This will complete the circuit with the top of the battery.

3. Put it all together.

Place the battery in the base of the metal body with the negative (−) side up. Slide the motor into the upper housing and position it so the straight conductor is inside the housing and the bent conductor touches the top of the battery. Use pliers to compress the housing and hold the motor in place, being careful not to crush it.

Use It

If everything checks out, it should already be running. Put it on a flat surface — it'll spin around and occasionally fall over. If it falls over more than occasionally, adjust the alignment of the base with pliers.

When your Dizzy Robot has had its fun, slide a small scrap of card between the top of the battery and the motor contact to turn the robot off.

Steve Hoefer makes things, solves problems, and is the main brain behind grathio.com.

YOU WILL NEED

Vibrating motor such as **#G16777** from **goldmine-elec.com**, **$1**
AG13 button cell battery
Needlenose pliers

0.008" sheet tin from a hobby, art, craft, or hardware store
Tinsnips or heavy-duty craft scissors

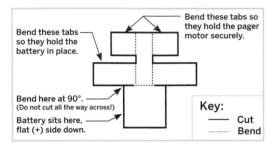

Bend these tabs so they hold the battery in place.

Bend these tabs so they hold the pager motor securely.

Bend here at 90°. (Do not cut all the way across!)

Battery sits here, flat (+) side down.

Key:
— Cut
····· Bend

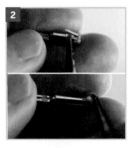

Photography and diagram by Steve Hoefer

Make: Projects

Get a bird's-eye view of your neighborhood with our helium balloon photography rig, designed by a NASA engineer. Then, back on Earth, make a stroboscope to take freeze-frame photographs of fast-moving objects. Finally, build an electromagnetic levitator that shoots rings of non-magnetic aluminum (and learn why it works).

Helium Balloon Imaging "Satellite"
80

Stroboscope
90

Electromagnetic Aluminum Levitator
100

HELIUM BALLOON IMAGING "SATELLITE"

By Jim Newell

INFLATION WATCH

Snap aerial photos from 300' up by suspending a hacked drugstore camera from 3 tethered helium balloons.

The first time I saw a satellite photo of my house on Google Earth, I expressed shock at the "Big Brother" implications of an all-seeing, commercial eye-in-the-sky. But meanwhile, I was also secretly disappointed with the picture quality and clarity because (Orwellian angst aside) I needed better overhead images for my own use — to help me lay out a new driveway and complete a birds-eye-view CAD drawing of our lot.

So I decided to design and fabricate a simple helium balloon "satellite" camera platform, tethered to the ground for ease of control and retrieval, and dedicated to a single purpose: to capture aerial images of my house and surroundings.

Here's how I completed this project using inexpensive and readily available components — helium balloons on a nylon kite string, a drugstore camera perched on a platform made out of an old CD, and a PICAXE microcontroller housed in an empty pill bottle.

Photograph by Sam Murphy

Set up: p.83   Make it: p.84   Use it: p.89

Jim Newell (jamesmnewell.com) has degrees in mechanical engineering, physics, and business administration, and has worked in the aerospace industry for the past 28 years. With interests including electronics and home automation, he's also an avid guitar player and singer/songwriter.

Balloon Cam: How It Works

A tiny $3 microcontroller chip tells your aerial camera when to snap.

A cheap, lightweight (A) digital camera, modified for electronic shutter triggering, snaps photos. The camera's lens aims down through the center hole of a spare (B) CD (or DVD), which serves as the camera's support platform. The CD is suspended from the bottom of a (C) pill bottle that holds the trigger board and battery. The (D) trigger board uses a tiny PICAXE-08M (E) microprocessor that's programmed (in BASIC) to send repeating pulses from one of its output pins. The pulses are applied to a (F) reed relay, which opens and closes the connection between the camera's shutter control contacts, repeatedly triggering the shutter. Power for the board comes from a (G) 9-volt battery via a (H) voltage regulator that drops the circuit's voltage to 5V. The compiled microprocessor code is uploaded to the PICAXE through a (I) programming header connected to a computer.

Calculating the Number of Balloons Needed

Total weight = weight of payload + weight of tether at max height = ½lb + ¼lb = ¾lb

Lift from helium at sea level (approx.) = 0.067lbs per cubic foot (ft³)

Volume of helium to lift ¾ pound = 0.75/0.067 = 11.2ft³

Volume of a sphere = $\frac{4}{3} \times \pi r^3$

Balloon radius needed (single balloon) = $(\frac{3}{4} \times 11.2 / \pi)^{1/3}$ = 1.4'

To account for the added weight and helium backpressure from the balloon itself, ensure sufficient lift for multiple flights, and add redundancy, I tripled this minimum figure and used three 3'-diameter balloons.

Illustration by Tim Lillis

SET UP.

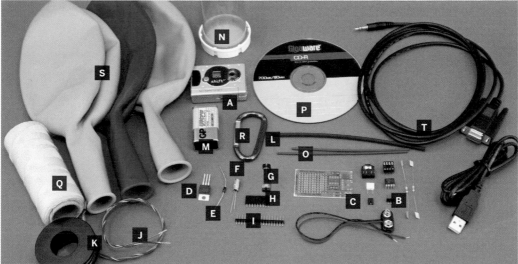

MATERIALS

[A] Digital camera, mini "keychain" $10–$15 from chain drugstores; brands tested were Shift3, Aries

[B] PICAXE-08M micro-controller SparkFun Electronics item #COM-08308 (sparkfun.com), $3

[C] PICAXE 8-Pin Proto Kit SparkFun #DEV-08321, $4

[D] Voltage regulator, 5V, TO-220 package, LM2940T or NTE1951 Digi-Key #LM2940T-5.0-ND (digikey.com), $2; or Mouser Electronics #526-NTE1951 (mouser.com), $4

[E] Diode, 1N4001 RadioShack #276-1101, (radioshack.com) $1

[F] Electrolytic capacitor, 22µF RadioShack #272-1026, $1

[G] Relay switch, 5V reed aka reed relay. RadioShack #275-0232, $3. RadioShack sells 2 versions of this under the same part number; the cylindrical package will work, but the box-shaped one is too large to fit.

[H] Connector header, female, 9×1, 0.1" spacing Digi-Key #S7042-ND, $1

[I] Breakaway headers, male SparkFun #PRT-00116, $2

[J] Electrical wire, 18–20 gauge, insulated

[K] Solder

[L] Heat-shrink tubing or electrical tape

[M] Battery, 9-volt

[N] Pill bottle with child-proof cap large enough to hold microcontroller and battery

[O] Solid wire, 12 gauge

[P] Compact disc or DVD

[Q] Kite string, 150lb test, I used Dacron Archline from Conwin (conwinonline.com), 200yds for $15.

[R] Small carabiner

[S] Balloons, 3' diameter (3) Try BalloonsFast (balloonsfast.com/ 36inchlatexballoons) $4 each

[NOT SHOWN]

Helium, 42 cubic feet You can have balloons filled at a balloon supplier or a super-market, but the supermarket balloon person will probably not be happy with you using so much helium.

Duct tape

Double-sided foam tape

TOOLS

[T] PICAXE USB pro-gramming cable PICAXE part #AXE027, SparkFun #PGM-08312, $26. Or, with a Windows or Linux PC, you can save money by buying the $15 PICAXE-08M Starter Pack, SparkFun #DEV-08323, which includes the PICAXE 8-Pin Proto Kit (see Materials list) and a serial programming cable rather than USB.

[NOT SHOWN]

Computer with internet connection

Soldering equipment

Wire strippers

Wire cutters

Phillips screwdriver, small to unscrew camera case

Flathead screwdriver, small to pry open camera body

Tweezers

Multimeter or ohmmeter

Drill and drill bits: ⅛", ¼"

Pliers

Scissors

Helping hands with magnifier (optional)

Tabletop vise (optional)

MAKE IT.

BUILD YOUR BALLOON-SAT

START ⋙   Time: 1–2 Days Complexity: Moderate

1. MAKE THE TRIGGER BOARD

For all connections, refer to the project schematic seen here (downloadable at makezine.com/24/ballooncam). To stabilize voltage regulator operation, I added an additional 22µF electrolytic capacitor across the power leads, in parallel with the Proto Kit's included 100nF cap.

1a. Assemble the PICAXE Proto Kit, soldering the following components at the locations indicated on the printed circuit board (PCB): 8-pin IC socket, stereo download socket, 3-pin header, 10kΩ resistor, 22kΩ resistor, and 100nF capacitor. These components are small, so a helping hand with magnifier or a tabletop vise will come in handy. Don't connect the battery clip yet.

Once all parts are in place, carefully place the microprocessor chip in its socket, with pin 1 (indicated by the notch) pointing away from the prototyping area (or you can place it later; see Step 1e). Also, move the jumper on the 3-pin header to the PROG side to enable it for programming.

1b. Thread the leads of the battery clip through the 2 holes in the Proto Board PCB, and solder the black wire into place on the bottom of the board.

1c. Insert the voltage regulator through 3 holes near the center of the PCB. Run the red wire from the battery clip across the top of the PCB and connect it to the regulator's input pin (indicated by a dot). Using a wire jumper, connect the middle (ground) pin of the regulator to the black wire from the battery clip. Use another wire jumper to connect the LM2940 output pin to the PCB, at the location marked "RED," where the red wire from the battery clip would normally attach.

Photography by Jim Newell

1d. Insert the 22µF capacitor through the 2 PCB holes indicated by (+) and (−), and solder it in place. Be sure to watch polarity; the stripe on the cap goes on the (−) side.

1e. Insert the relay into the PCB at the forward edge so that the single switch terminal dangles off the side of the board and the other 3 pins run through holes. The bottom left pin should run through the hole that's 3 up from the bottom of the board and 2 from the left.

1f. Insert the 1N4001 diode into the through-holes that connect to the relay coil terminals so that its body drapes over the top of the relay. This diode protects the PICAXE from back electromotive force when the relay is de-energized. Using jumper wires, connect one of the relay coil terminals to the PCB ground pin and the other to PICAXE output 2 (pin 5 on the chip).

1g. Solder about 1' of 18- to 20-gauge wire to each of the relay switch terminals. Use wire cutters to cut a 2-pin length from the male breakaway headers, and solder the 2 pins to the wires' other ends. Cover the solder joints with heat-shrink tubing or electrical tape to prevent shorts and to add strength.

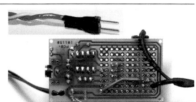

2. PROGRAM THE TRIGGERING

2a. Download and install Revolution Education's free AXEpad software from rev-ed.co.uk/picaxe.

2b. Download the BASIC file *Camera_Timer.bas* from makezine.com/24/ballooncam, then open it up in AXEpad.

This simple, 14-line routine waits 20 seconds from the time of initial power-up to give time to replace the pill bottle cap, takes one picture to confirm that it's running, waits another 20 seconds to let the balloon rise, then begins snapping pictures every 2 seconds. You can modify this to suit your needs.

2c. To load this code into your Proto Board, connect the PICAXE programming cable between your computer and the board's programming jack, then click the Program button in AXEpad, in the upper right.

3. MODIFY THE CAMERA

The specifics of this step will depend on the camera you use, but it's a simple mod, and readers with basic electronics skills should have no problem. The camera shown here is Shift3 brand and was purchased for $11 at a Rite Aid pharmacy.

3a. Remove the stick-on label from the front of the camera (or the side, for the Aries camera) to reveal a screw that holds the case together.

3b. Remove the screw with a small Phillips screwdriver.

3c. Gently pry the camera shell open using a flathead screwdriver.

3d. Remove the 2 screws holding down the circuit board, and also unscrew the keychain clip, which we don't need (with the Aries camera, remove 3 screws to detach the board).

3e. Turn the board over so that the lens is visible. Handling the board by the edges only, and without touching any parts, use tweezers to remove the black potting material from around the shutter switch, exposing solder terminals at its base.

3f. Solder wire leads to the 2 newly exposed switch terminals.

3g. To make room for the shutter switch wires to exit the case, use pliers to cut a hole in the plastic on the side opposite the lens (with the Aries camera, instead of cutting the case you can remove the pop-up viewfinder lens assembly and route the wires out of its hole.)

3h. Replace the board and screw it back into the case, routing the shutter switch wires out, and reassemble the case.

3i. Cut a pair of adjacent female connector headers, and solder one of the wires to each. This will connect to the male header pair from the trigger board.

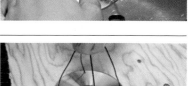

4. ASSEMBLE THE SATELLITE STRUCTURE

4a. Drill 4 equidistant ⅛" holes around the bottom of a pill bottle large enough to hold the microcontroller board and battery (about 2" in diameter and 4" tall). Drill another hole through the center.

4b. Insert 6" lengths of stiff 12-gauge solid wire into the 4 perimeter holes and extend them downward from the bottom of the bottle. Inside the bottle, bend the tops of the wires so they stay in place when you pull the wires from below.

4c. Mark and drill 4 uniformly spaced ⅛" holes around the periphery of a spare CD or DVD.

4d. Thread the 4 wire standoffs from the bottom of the pill bottle through the 4 holes on the CD, bend them to lay flat underneath the surface of the CD, and secure them in place with duct tape, or by twisting them up and around.

4e. Grab the spool of kite string, and route the free end up through the center hole in the CD and through the center hole in the bottom of the pill bottle. Tie it off to the 12-gauge wires inside the bottle.

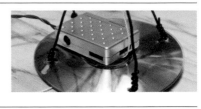

5. ATTACH THE CAMERA AND BALLOONS

5a. Mount the modified camera to the top surface of the CD with double-sided foam tape so that the lens of the camera looks down through the hole in the center of the CD.

5b. Insert the PICAXE board and 9-volt battery into the pill bottle, but don't connect the battery yet (powering up the board will start the program running).

5c. Drill a ¼" hole in the center of the pill bottle cap and 4 more small holes around the periphery, uniformly spaced and as close to the outside diameter as possible.

5d. Route 4 pieces of kite string, 8"–12" each, through the perimeter holes and tie them all together securely inside the cap. Then tie together the other ends of the 4 strings, extended from the top side of the cap, to a small carabiner.

5e. Inflate the balloons with helium, tie each one off with a knot, and tie on a 1'–2' piece of kite string. Tie the balloons together to form a tight group, and tie them all to the carabiner. When finished, you'll have the balloon group attached to the cap of the pill bottle, and because the balloons attach to the satellite only through this cap, it must be of the childproof variety to make sure it stays on securely.

5f. Finally, attach the bottle cap, routing the camera connector wire through the center hole. You're ready to fly!

FINISH ☒

NOW GO USE IT »

USE IT.

NUDE SUNBATHERS BEWARE!

BALLOON-SAT OPERATION

Once the balloon-satellite is fully assembled you're ready to launch. Here's how:

1. Plug the headers together from the trigger board and the camera, and turn the camera on.

2. Unscrew the top of the pill bottle, being careful not to let it fly away. Inside the bottle, connect the 9-volt battery to the Proto Board battery cable, and quickly screw the top back on.

3. Let the satellite go — up, up, and away!

If you look carefully at my photos, you can see the kite string along with a knot I had to tie due to some poor planning. I hope that readers can plan their string routing a little better than I did and keep the images knot-free.

ENHANCEMENTS

You can extend the trigger board for greater functionality by adding a second relay to switch the camera on and off. That way the keychain camera could turn on once it reaches altitude and start capturing aerial home movies in video mode. Not even Google Earth can compete with that!

STROBOSCOPE
By Nicole Catrett and Walter Kitundu

PLAY WITH SPACE AND TIME

Make a mechanical strobe with a toy motor and construction paper, pair it with a digital SLR camera, and take stunning photographs of objects in motion.

We were inspired to play with stroboscopic photography after seeing photographs taken by 19th-century French scientist Étienne-Jules Marey. In the 1880s, Marey invented a camera with a rotating shutter that could capture multiple images on a single photographic plate. He used this camera to study locomotion in humans, animals, birds, sea creatures, and insects.

Marey used clockwork mechanisms and photographic plates for his contraption, but you can make a much simpler version with a slotted paper disk, a toy motor, and a digital camera. The camera is set to take long exposures while the slotted disk spins in front of its lens. Each time the slot spins past the lens, the camera gets a glimpse of your subject and adds another layer to the image. The resulting photograph is a record of your subject moving through space and time, and these images often reveal beautiful patterns that would otherwise be invisible to us.

Set up: p.93   **Make it: p.94**   **Use it: p.99**

Nicole Catrett is an exhibit developer at the Exploratorium in San Francisco. You can play with her stroboscope exhibit there and see more of her work at nicolecatrett.com. **Walter Kitundu** is an artist and bird photographer. You can see more of his work at kitundu.com.

Photography by Walter Kitundu

Freeze Frame

A spinning disk with a single slit lets your camera see serial glimpses of a moving subject — and record them all in a single image.

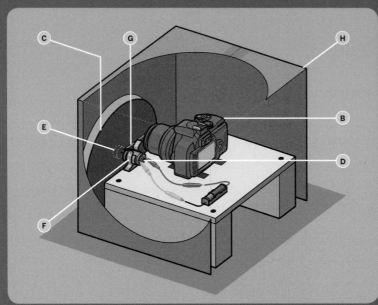

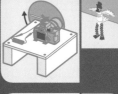

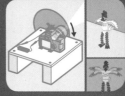

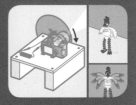

A The subject moves through space. Lights and a black background ensure that each successive image is sharp and distinct. (See page 98 for setup tips.)

B A camera is aimed at the subject.

C A black strobe disk covers the camera's lens, preventing light from entering.

D A battery-powered DC motor spins the strobe disk.

E A cork in the center of the disk makes it easy to mount and remove from the motor.

F The motor is mounted to a base with a broom-holder clip.

G A slit in the spinning strobe disk lets a bit of light into the lens as it passes over. When the camera is set to a long exposure time, each slit passage allows a new image to overlay onto the camera's image sensor.

H A cardboard camera hood reduces ambient light that can cloud the captured images.

Illustration by Rob Nance

SET UP.

MATERIALS

[A] Cork from a wine or champagne bottle

[B] Battery holder, 1xAA and battery RadioShack (radioshack.com) part #270-401, $1

[C] Motor, small, 1.5V–3V DC RadioShack #273-223, $3

[D] Alligator test leads (2) RadioShack #278-001, $7 for a set of 4

[E] Foam sheet ("Foamies"), 2mm×9"×12" Craft Supplies For Less #F2BB10 (craftsuppliesforless.com), $4 for a pack of 10

[F] Broom-holder spring clip with mounting screw McMaster-Carr #1722A43 (mcmaster.com), $11 for a pack of 10. You can also use a zip tie.

[G] Zip tie

[H] Digital SLR camera with manual focus that can take long exposures Some good options are the Nikon D40, D3000, and D5000, and the Canon EOS Rebel Ti, XTi, XS, or XT. You might be able to use a point-and-shoot camera, but it needs to have a manual focus and exposure mode.

[I] Plywood, ½" thick, 10"×10"

[J] Lumber, 2×4 , 10" lengths (2)

[K] Wood screws, 1¼" (4)

[L] Black fabric, enough to work as a backdrop A queen sheet is a good size, and black cotton works well. Avoid anything shiny or reflective. You could also use matte-finish black paint for your backdrop.

[M] Clamp lights (2 or more) Home Depot #CE-303PDQ (homedepot.com), $13 each

[N] Velcro tape, 3½" strips OfficeMax #09015086 (officemax.com), $4 for a pack of 10

[O] Construction paper or cardstock, black (1 sheet)

[NOT SHOWN]

Cardboard box approximately 12" square

TOOLS

[P] Drawing compass and pencil

[Q] Ruler or calipers

[R] X-Acto knife

[S] Scissors

[T] Hot glue gun

[U] Sewing needle, large

[V] Phillips screwdriver

[NOT SHOWN]

Drill and drill bits

MAKE IT.

BUILD YOUR STROBOSCOPE

START »» Time: **2–3 Hours** Complexity: **Easy**

1. MAKE THE STROBE DISK

1a. Use a compass to draw an 8½"-diameter circle on black construction paper or cardstock; apply enough pressure with the compass point to mark the circle's center.

1b. Measure the diameter of your cork. Use the compass to draw a circle of that diameter at the center of the larger circle. This will come in handy later when you mount the strobe disk on the motor shaft.

1c. For the strobe disk slot, use a ruler to draw a line from the center mark to the outer edge of the big circle. Then draw a second line ⅛" away from and parallel to the first line.

1d. Make a perpendicular mark across the 2 slot lines ½" in from the edge of the big circle. The slot will run from this mark to the closest edge of the inner circle.

1e. Cut out the slot with an X-Acto knife, using the ruler as a guide to make the cuts clean and straight.

1f. Use scissors to cut out the large circle.

2. ATTACH THE DISK TO THE MOTOR

2a. Use an X-Acto knife to slice a round section from the cork, ½" thick.

2b. Hot-glue the flat side of the cork section to the center of the strobe disk, using the small center circle you drew as a guide.

2c. Use a large sewing needle to pierce the center of the cork through the disk. This will make a pathway for the motor shaft.

2d. Push the strobe disk and cork onto the motor shaft, with the cork facing out. This should be a snug fit.

2e. To test the motor and strobe disk, use alligator test leads to connect the toy motor to your AA battery holder. The strobe disk should begin to spin.

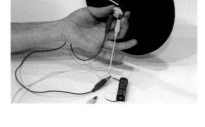

You now have the fundamental parts of your stroboscope. Peer through the contraption at anything moving, and the scene will turn into an old-time movie. Watching your friends dance will be a whole new syncopated experience. Also try looking at a vibrating guitar string or a stream of water.

3. BUILD THE BASE

3a. Cut a 10" square piece of plywood for the base platform. Position a broom-holder clip (screw-hole down) with its long side parallel to an edge of the plywood, spaced approximately ¼" in from the edge and 2" in from a corner. Screw the clip down tight. (If you're using a zip tie instead of a broom clip, see Step 3c.)

3b. Remove the strobe disk from the motor and set it aside. Wrap a thin foam sheet (cut to fit) around the motor and pop it into the broom holder, shaft pointing out. The foam keeps the motor from vibrating.

3c. (Alternative) If you don't have a broom holder, you can use a zip tie. Drill 2 parallel holes (each large enough to accommodate the diameter of your zip tie) ¼" in from the edge of the plywood and 2" and 3" from the corner, respectively, and pass the zip tie through the holes and around the motor to secure it to the plywood.

3d. For the "legs" of the base, cut two 10" lengths of 2×4 lumber. Position the legs parallel to each other (upright on their long sides) about 7" apart, and place the plywood square on top so it lines up with the outside edges of the legs. Mark and drill pilot holes in the corners of the plywood square and legs, about ¾" in from each outside edge, and use the 4 wood screws to securely attach the base together.

3e. Put the strobe disk back on the motor shaft.

4. MOUNT THE CAMERA

4a. If your camera has a zoom lens, set it to the widest angle possible. Place your camera on the wooden base so that its lens points at the strobe disk, completely within the slot's path. To keep extra light from entering the lens, the disk and lens should be as close as possible without touching.

4b. Use a pencil to mark the location of the camera's body on the base. Set your camera aside.

4c. Adhere two 3½" strips of velcro tape (hook side) onto the wooden base in the camera body location, parallel to each other and perpendicular to the image plane. The velcro will hold the camera in place and let you make fine adjustments to its position.

4d. Place one strip of velcro (loop side) along the bottom of the camera body. Make sure you can still access the battery compartment.

4e. Position your camera on the base. If it tips forward, slice a round section of cork to place flat under the lens for support; dry-test the fit, then remove the camera and hot-glue the cork to the base.

4f. Put your camera back in position, and locate a place for your battery holder a couple of inches behind your camera and opposite the motor. Remove the battery and use hot glue to attach the battery holder to the base.

4g. Replace the battery and reattach all but one of the alligator test lead connections.

5. MAKE THE HOOD

5a. Find a cardboard box big enough to turn upside down and fit over your whole rig, with room to spare. Use an X-Acto knife to cut out the back wall so you can access the camera. Cut a round opening in the front of the box, to give your camera and strobe disk a clear line of sight.

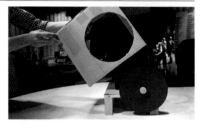

5b. Place the box over your rig to check the fit. Make sure the camera still has a clear view and that the slot in your strobe disk isn't obscured. Voilà — your stroboscope is complete!

GETTING SET UP
Backdrop, Subject, and Lighting
Good stroboscope photography requires a black backdrop, preferably fabric, which makes your subject show up clearly without being lost in a bright clutter of background noise. With a black background, light and brightly colored objects will "pop," while dark objects will disappear.

Your subject also needs to be well lit, or else it won't show up. Clamp lights work well and are easy to adjust. With the camera pointing straight toward the backdrop and your subject in between, place 2 clamp lights pointing in from the left and right, respectively, lighting up the subject rather than the backdrop.

You can also set up your stroboscope outside, using sunlight instead of clamp lights. As long as you have a black background and bright light, you're in business.

Camera Settings
To set up your camera, temporarily remove the strobe disk and focus your lens on the place where your subject will be moving. Make sure the focus remains set to manual. Here are some good typical settings you can experiment with for starters:

Shutter/exposure: Two seconds. With a too-short exposure, you won't see much happening in the image, but if it's too long, the image will be washed out.
Aperture: Set this to the lowest number possible to gather the maximum amount of light with each pass of the strobe disk slot.

FINISH X

NOW GO USE IT »

USE IT.

HAPPY TRAILS!

THE FUN PART

Now you can start to play with stroboscopic photography. Have a friend press the shutter while you try tossing or bouncing balls, juggling, releasing balloons, throwing sticks or paper airplanes, doing cartwheels, or dancing. Almost anything that moves is fun to photograph with a stroboscope. One of our favorite things to play with is string. Try twirling it in spirals or jumping rope.

Don't be discouraged if your first few images are out of focus or washed out. You can solve these issues by adjusting your camera settings, making sure the cardboard hood is in place, and adjusting or adding lights.

The camera will be focused at a single fixed distance, so it helps to mark the floor to remind yourself or your subject to stay in the plane of focus. This will instantly improve your images.

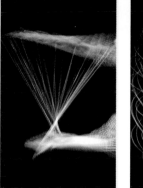

DISK VARIATIONS

Large, slower-moving subjects (like people) look better when the disk only has one slot. However, certain fast-moving subjects — such as thrown objects or vibrating strings — look better if you use a strobe disk with 2 or 3 slots, which doubles or triples the number of exposures per second. Follow Steps 1a–2c to make additional disks that have multiple slots, and make sure that the slots are spaced apart evenly, so the disk stays balanced while it spins.

ELECTROMAGNETIC ALUMINUM LEVITATOR

By Thomas R. Fox

POWERS OF INDUCTION

This simple AC-powered coil device uses magnetism to levitate aluminum rings and shoot them into the air — and aluminum isn't even magnetic!

About 13 years ago I learned of some military research into a satellite- and missile-defense device that would propel projectiles using Lenz's law, which governs the direction of electrical current induced by a changing magnetic field. I decided to make a little gadget based on the same principle for my kids, to get them interested in science and electronics. The gadget worked — both at levitating and shooting rings, and at interesting my kids (some of them, anyway).

This design is actually the third one I tried. I had no real information on what the military was up to, so I tested my own ideas. One iteration used wire wound around a steel pipe with a short piece of aluminum rod inside. (That one didn't work — trust me!)

Not only is this project physically exciting and intellectually stimulating, it's also quick, easy, and inexpensive to build. Except for the 200 feet of #18 magnet wire, you can purchase everything you need at a hardware store or scavenge it from common junk.

Set up: p.103   Make it: p.104   Use it: p.107

Tom Fox (magiclandelectronics.com) is the author of numerous articles about electronics, and of three books, most recently *Snowball Launchers, Giant-Pumpkin Growers, and Other Cool Contraptions*. He is also the workshop editor at *Boys' Quest* and *Fun for Kidz* magazines. Tom and his family own and operate Magicland Farms (magiclandfarms.com), a farm and roadside market in Michigan.

Photograph by Sam Murphy

Flying Rings

Electric current generates a magnetic field, and a changing magnetic field will induce electric current. By Lenz's law, the direction of induced current will oppose the direction of the changing field.

Using these principles, the Electromagnetic Aluminum Levitator converts ordinary AC power into 2 opposing magnetic fields — one from a stationary coil, the other from a ring-shaped projectile.

A The power strip supplies 120V AC to the device when its switch is turned on.

B The coil generates a strong, alternating-direction magnetic field from the AC power. A stand holds the coil steady.

C A time-delay 7A fuse prevents the coil from heating to dangerous, insulation-melting levels if the switch is left on. About 10A of current runs through the coil, but the fuse's time delay allows the current to be switched on for safe intervals of 4 seconds or less.

D A steel or iron rod extends the magnetic field created by the coil.

E The coil's magnetic field induces an electric current in the aluminum ring that runs opposite the direction of current in the coil. This creates a magnetic field from the ring that opposes the field from the coil, following the right-hand rule.

F The opposing magnetic fields between the coil and aluminum ring cause the ring to levitate or to shoot up into the air.

AC Not DC

Connecting direct current (DC) to the levitator's coil would turn it into an electromagnet — but that wouldn't affect an aluminum ring, because aluminum isn't magnetic. Alternating current (AC), in contrast, creates a constantly changing magnetic field that induces current in the conductive aluminum ring — the same way that moving coils toward and away from stationary magnets induces current inside a generator.

Following Lenz's law, the current in the ring runs opposite its direction in the coil, which causes repulsion between the two. The force is strongest next to the coil, so the ring will fall to an equilibrium point and levitate when dropped over a powered coil, but if you put the ring on first and then apply power, it shoots up.

The levitator will also act on rings of copper and other conductive materials, but aluminum works far better because of its light weight.

Frequency Response

A changing magnetic field induces a force on electrons proportional to its rate of change, so the higher the frequency of AC you apply to the levitator's coil, the greater the voltage (and opposing magnetic force) you'll induce in the ring. But higher frequency also increases a conductor's impedance, which dissipates energy and reduces current. So there's some happy-medium AC frequency that would provide optimal levitation — it probably isn't standard 60Hz, but it might be close.

The coil wire has a DC resistance of only about 1.5Ω. So by Ohm's law ($I=V/R$), you might think that a scary 80A will flow through it when you connect it to 120V AC. But coiling the wire around a metal rod creates an inductor with an AC impedance of about 12Ω at 60Hz, permitting a much lower (but still dangerous) 10A.

Illustration by Damien Scogin

SET UP.

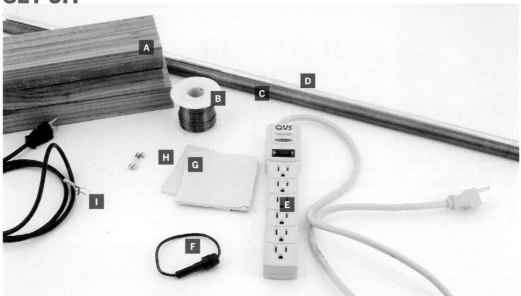

MATERIALS

**[A] 2×4 lumber
(1½"×3½"), 3' long**

**[B] Magnet wire, #18, 1lb
spool (199')** Mouser
Electronics #566-8075
(mouser.com)

**[C] Iron or steel rod (not
stainless), ¾" diameter,
2'–3'** I used a cold-rolled
steel rod, but purified iron
would work best.

**[D] Aluminum tubing,
inside diameter ¹³⁄₁₆"–1",
at least 1' long** Hardware
stores sell this, but I cut
mine from the handle of an
old snow rake I used before
global warming set in. Fishing
and pool nets also frequently
use aluminum tubing, and
check out yard and garage
sales for other sources of
cheap aluminum tubing.

**[E] Power strip with switch
and fuse or circuit breaker**
UL listed and rated at 12A
or more

[F] Fuse holder RadioShack
#270-739 (radioshack.com)

**[G] ¼" plywood, 4"×4"
squares (2)** or similar
material. I used scrap
epoxy-glass printed circuit
boards (PCBs).

**[H] Time-delay fuses, 7A,
¼"×1¼" (1+)** It helps to
have some spares handy.

**[I] Heavy-duty grounded
extension power cord at
least 15'–25' long** rated at
13A or more

[NOT SHOWN]

**Wire, #14 or #16,
insulated, 3'**

Hose clamp, small

Electrical tape

Epoxy cement

**Wood screws, #8, 2½"
long (4)**

**Brightly colored tape
(optional)**

**Heat-shrink tubing
(optional)**

TOOLS

Hacksaw

**Drill and drill bits: ⁵⁄₆₄",
¹¹⁄₁₆", ¾", countersink**

Awl or thin pencil

Screwdriver

Soldering tools and solder

Small knife or sandpaper

Multimeter or ohmmeter
that measures down
to ¹⁄₁₀Ω

Photograph by Ed Troxell

MAKE IT.

BUILD YOUR ELECTROMAGNETIC ALUMINUM LEVITATOR

START ⋙   Time: **An Afternoon** Complexity: **Easy**

1. MAKE THE STAND

Cut the 2×4 into three 1' lengths. Drill a ¾" hole centered through one of them.

Attach the ¾" drilled wood piece crosswise over the other 2 pieces to make an "H" shape. First, mark and drill two ¹¹⁄₆₄" holes staggered at each end of the center piece using a countersink (see figure at right). Next, with an awl or thin pencil, mark pilot hole locations on the middles of the other pieces through the countersunk holes. Drill ⁵⁄₆₄" pilot holes, line them up with the holes on the center piece, and screw the stand together with four 2½" #8 screws.

Attach the 2×4 boards together with four 2½" #8 wood screws

Drill ¾" hole in center of top board

2. WIND THE COIL

2a. Cut the plywood or PCB material into two 4" squares and drill a ¾" hole in the center of each.

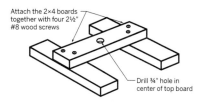

2b. Jam the 2'–3' metal rod through the hole in the stand so it protrudes 1" from the bottom, and slide one of the 4" squares down the rod from the top, resting it on the stand (about 3" from the end of the rod). Use epoxy to cement the square in place on the rod — then, using your wire spool for spacing and temporary support, slide the second 4" square onto the rod and epoxy it in place, about 2½" up from the first. Let the epoxy cure.

Photography by Ed Troxell

2c. Wrap electrical tape around the rod to completely cover the area between the 2 squares. This adds extra insulation between the coil and rod, for safety.

2d. Leaving 2"–3" of length free at the start and finish of the coil, tightly wind the full spool of #18 magnet wire (approximately 200') around the rod between the 2 squares. These lengths of wire will connect the coil to the 120V AC power source later. Once the coil is done, use electrical tape to prevent the wire from unraveling.

3. CUT THE RINGS

Use a hacksaw or similar tool to cut several rings from the aluminum tubing. They should be at least 1" long, but you can cut different sizes to experiment. You'll need a whole bunch; the rings can shoot up so high that you might lose some!

You can try cutting rings from other materials, although steel or iron tubing won't work well because it's not only heavier and has a lower electrical conductivity than aluminum, it's also inherently magnetic and will be attracted to the coil during those 2 instances in every cycle when the voltage isn't changing. Copper tubing conducts electricity even better than aluminum, but it's also heavier than aluminum and doesn't react physically (due to inertia) as strongly to the changing magnetic field as aluminum.

Rings cut from any type of tubing should have an inside diameter of 13/16"–1".

Optional: To make the rings easier to see and photograph, wrap the outsides with brightly colored tape.

4. WIRE IT TOGETHER

This project has only 3 electrical parts: the plug (P1 in the schematic at right), the coil (L1), and the fuse (F1). The time-delay 7A fuse is required to keep the coil from heating to possibly dangerous levels if you leave the power-strip switch on.

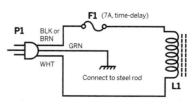

4a. Use a knife or sandpaper to scrape 1" of enamel insulation off each end of the coil's magnet wire.

4b. Cut off the socket end of the heavy-duty power cord, and strip ½" of insulation from each of its 3 wires. Use a multimeter to confirm that the wires are color-coded correctly: with the cord's plug pointing away from you, and with the round prong (ground) at the bottom, the white wire should connect to the left, usually larger (neutral) flat tab; the black or brown wire to the right, usually smaller (hot) flat tab; and the green wire to ground.

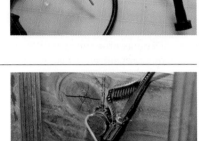

4c. Solder the cord's hot (black/brown) wire to one side of the fuse holder, and solder the neutral (white) wire to one end of the coil. Use heat-shrink tubing or electrical tape to insulate this and all other solder connections.

4d. To ground the metal rod, strip an additional 3" of insulation off the green wire, wrap it around the bottom of the rod, and secure it with a hose clamp. If the rod is corroded, sand or scrape the contact area first to ensure a good connection.

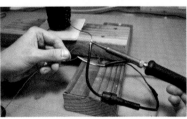

4e. Solder a short length of #14 or #16 insulated wire to connect the free end of the coil to the free end of the fuse holder.

5. PRELIMINARY TESTING

5a. Connect the multimeter's leads to measure resistance between the 2 flat tabs (hot and neutral) on the plug end of the power cord. It should measure between 1.1Ω and 2Ω. Leave one test lead connected to its tab and apply the other one to the center ground prong; it should register infinite ohms, no connection. Move the ground probe from the prong and touch it to the outside of the coil (the wire's enamel coating) and the metal rod. It should measure infinite ohms for both of these. If the multimeter detects *any* conductivity (less than infinite ohms), **DO NOT apply power to the device until the problem has been resolved.**

5b. To make sure the rod is grounded, use the multimeter to check continuity from the plug's ground prong to the metal rod. If this checks out too, you're done!

⚠ **SAFETY WARNING: The rod must be grounded!** This ensures that the circuit breaker will trip if there's a short through any insulation, and that touching the rod won't cause a shock.

FINISH ⊠

NOW GO USE IT ≫

Photograph by Thomas R. Fox (4d)

USE IT.

LET'S FLING SOME RINGS

PLUGGING IN

First, make sure any outlet you plug the power strip into is properly grounded. Test the AC voltage between the smaller, right (hot) slot and the bottom, rounded (ground) slot. It should read 110V–120V AC. If it's outdoors, the outlet must be a UL-listed GFCI (ground-fault circuit interrupter) type. These outlets have test and reset buttons, and are often used where moisture might cause short circuits.

⚠️ **CAUTION: When setting up, place the stand a healthy distance away from the switch-operated power strip. This is why you used an extension cord at least 15' long. Also, place the stand on a high table or other flat, raised surface so that no one can look down over the rod and accidentally get hit in the eye.**

FLOATING AND FLYING

To make the ring levitate, switch the power strip on first, then immediately slip the ring over the rod using one hand (keep the other behind your back). This is easier with 2 people (and safer and more fun). Notice that the ring stops and hovers instead of falling. The ring is levitating! Then shut off the power immediately, or else the fuse will blow within a few seconds and need to be replaced.

For more fun, try shooting the ring upward. For safety's sake in your first attempt, wrap the ring with 2 layers of electrical tape. This will weigh it down so it won't shoot very high. Put the ring on the rod and let it slide down to the coil. Step back about 10', then turn on the power strip. The ring should shoot up enough to come off the rod. If you want it to go higher, remove some or all of the ring's tape.

FASTER AND HIGHER!

This is a demonstration project. While its operation is impressive, you can shoot the ring higher and faster if you make a few changes. For all these enhancements, the most important precaution is to locate the coil and rod even farther away from

the switch, yourself, and any observers. To do this, use a 50' extension cord in the build, or add it in between the power strip and the levitator plug. Any enhancements are for shooting only — not the levitating trick. No one should get close to the live rod.

One improvement is to use a rod made from special purified iron instead of the cold-rolled steel rod. Second, you can try using heavier-walled aluminum for the ring, and experiment with different lengths. Third, you can make a larger coil with more turns.

With any such "enhanced levitator," make sure you take the same precautions used by model rocket hobbyists. Following are some rules adapted from the National Association of Rocketry's safety code (nar.org/NARmrsc.html).

1. Use a countdown before launch, and make sure any guests are paying attention and are a safe distance away at the time of launch.

2. Place the levitator on a table or platform, and aim the rod close to vertical (within 30°). Never point the rod at anyone!

3. Test the levitator first outside in an open area that exceeds a 50' radius. Locate it at a safe distance from overhead power lines.

You may notice that the ring shoots up faster at times, with the least impressive performances generally occurring while you're showing someone you want to impress. The difference in heights depends on the instantaneous AC voltage, which in a typical house can vary from below 110V up to 120V.

⚠️ **SAFETY WARNING: Do NOT try to shoot the ring to the moon by hooking the coil up to a voltage source higher than 120V AC. For example, do not try this project with 240V AC. Even if the magnet wire's insulation holds up, you can start a fire or cause serious injury. Leave higher-voltage experimentation to the military or MythBusters!**

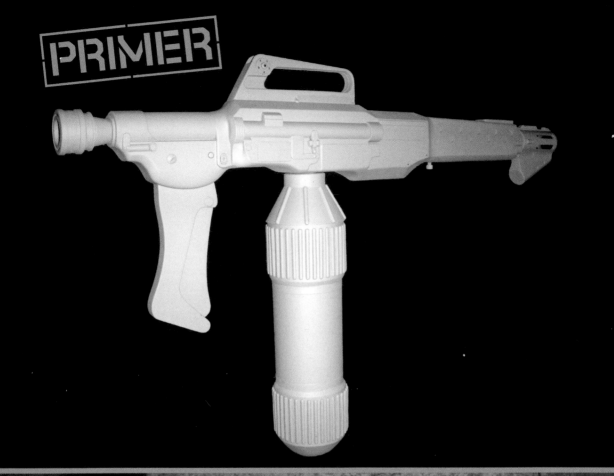

HARD-SHELL MOLDS

The science behind the art of prop making.

By Adam Savage

Silicone block molds are fine for making casts of small objects. But for larger items, like this 3-foot multi-piece prop gun (Figure A, page 111), you'd need hundreds of dollars' worth of silicone to make a block mold.

An excellent and inexpensive solution is to use a thin layer (or "blanket") of silicone that's keyed to a hard-shell or "mother" mold. It's a multi-step process, but it yields great results for the cost-conscious moldmaker. It also makes much lighter molds, which are easier to move around.

With this type of moldmaking, you're basically sculpting the 2 sides of your mold, taking into account the forces involved in the pouring and casting of the part. It takes a while, but if you get good at hard-shell moldmaking, you can cast just about anything, no matter how big.

GETTING STARTED

You'll need liquid silicone rubber, water-based clay, stone plaster, and a few other materials and tools; a complete list is at makezine.com/24/primer. The silicone is poured into 2 halves, and each half registers into its respective mother mold. Before doing anything else, take a permanent marker and draw a parting line down the exact middle of the original part, marking 2 symmetrical halves. This is the line you'll sculpt everything to — even with simple objects like this, every mold maker I know does it, and you should too.

1. PRE-FILL ANY VOIDS.

Before we make the mold itself, we need to make your original model mold-worthy. To prep it, you'll need to pre-fill up any small or difficult-to-access voids in the model, such as the hole in the bottom of our prop gun body, which you can see in Figure B (circled). This is a void on the model that has threads to join to another part of the model.

I knew that if I poured that void in 2 parts and from the side, I'd end up with air bubbles galore. So I pre-filled it with a plug made of blue silicone. The plug should stick to the blanket of silicone that I pour later, and become a nice part of the mold. Prior to pouring the silicone, you'll need to clean the plugs thoroughly with mineral spirits to make sure that they'll stick.

2. MAKE A FOAMCORE "TABLE."

Next, we'll make a table out of foamcore on which to sculpt the first half of the mold. Use a piece of foamcore that's large enough to extend at least 8"–10" beyond the borders of the part. (A common mistake people make with these molds is not giving themselves enough surface area to work on.)

Place the model on its side with the parting line parallel with the foamcore and trace around it with a pencil, as close to the model as possible. Cut the model shape out of the foamcore and discard.

Secure the model on a sturdy work table (oil-based clay is great for this step), making sure the parting line is as level as possible.

Now, we need to position the foamcore so that it sits about ¼" below the centerline of the model (Figure B). We'll be creating a clay dam to use as a reservoir for the silicone, and that ¼" allows for the thickness of the clay to come right to the middle parting line. Here's how to do it:

A. Measure from the parting line to the table. Let's say it's 2".

B. Subtract ¼" for your clay dam, which gives us 1¾".

C. Subtract the thickness of the foamcore (³⁄₁₆"), which takes us to 1⁹⁄₁₆".

D. Cut a few dozen little strips of foamcore 3"–4" long to this exact height, in our case 1⁹⁄₁₆".

E. Place the model on the table, and spread the upright strips around it evenly, covering an area the size of the cut-out foamcore piece.

F. Use hot glue to adhere the upright chunks to the table and to the large foamcore piece, which should be placed on top of them to create a nice solid foamcore surface.

3. MAKE A FUNNEL FOR POURING THE CASTING RESIN.

Now you need to decide where the top of your mold will be, into which you'll pour the casting resin. It should be located at the perimeter of the model where it comes to a steep point, so bubbles will surface and pop in a small area. Because you'll be pouring into this part, it's a good idea to add a piece to your foamcore table that gives you a nice clean surface for pouring the resin into the mold.

And you'll need a vent nearby to give the air you're displacing with the casting resin somewhere to go (other than back through the hole you're pouring from). This helps eliminate bubbles, and makes pouring the mold a cleaner, less splash-prone process. In Figure C, you can see the funnel-shaped pouring gate and next to it a smaller vent.

4. MAKE THE CLAY DAM.

Since you want the clay to be a uniform thickness, a rolling pin is ideal for the job. I use water-based clay, the kind used for pottery, because oil-based clay and silicone don't always get along. Flatten the clay with your hands, and when it's a little more than ¼" thick, place the clay pieces between 2 rails

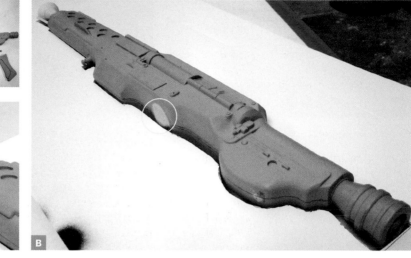

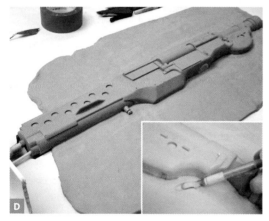

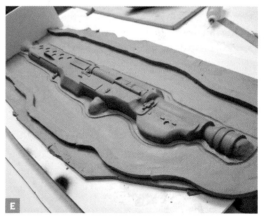

(wood will do) that are exactly ¼" high, positioned perpendicular to the rolling pin and close enough together that they're underneath its rolling surface. Roll the clay to a ¼" thickness — use a squirt bottle of water to keep it from getting sticky.

Figure D shows the clay dam, mostly laid out onto the foam board. It goes on in pieces, which you can join together with your fingers. Use a clay tool to bring the clay dam right up to the model. The line where the clay meets the model should be very smooth, perpendicular, and have no gaps. The cleaner this area is, the easier it will be to get good castings.

After finessing the clay dam, clean all clay residue off the model with a damp brush (otherwise it will be cast in when you pour your silicone!).

5. MAKE THE BARRIER KEY AND BORDER.

Use a wire-loop clay tool to make the first of the "barrier keys" (or more accurately, registration topography) that will hold the 2 halves of this mold perfectly aligned (Figure D, inset). Keys for molds come in many shapes and sizes, but for large molds, a key that runs all the way around the part helps prevent the resin from leaking out. I usually go a bit deeper than a half-circle's depth, then clean the edges with a wet soft paintbrush.

Figure E shows the completed barrier key around the part, and a raised clay border (made using the rolling pin method described in Step 4) built up all the way around the part, with spacing of about 3".

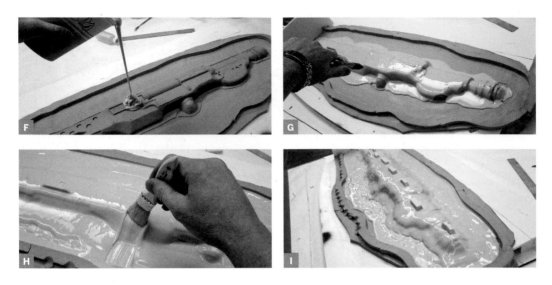

This provides enough space between the border and the model to add some plastic hemispherical keys to help the mold halves align ("register") properly; we'll discuss this in Step 9.

The clay bump you see in the middle is built around the pre-filled silicone plug I pointed out before. This clay plug will leave a void when I pour the first silicone blanket. That way, when I remove the clay plug and pour the second silicone blanket, the void will lead my silicone right to that pre-filled plug.

Also note the clay plug in the gun barrel. Unlike the other plug, I didn't pre-fill this shallow barrel with silicone. I didn't need to because it's large, easy to get bubbles out of, and has no threading. But the completed first blanket will have a trough that leads the silicone right into the barrel.

6. POUR THE FIRST BLANKET.

The first layer of silicone is the most important one, because it's what grabs all the detail from your model. To avoid bubbles, pour slowly from one location and from high up, letting the silicone drift slowly into the detail on the model (Figure F). You can use accelerator in the silicone (or use more kicker) to make it kick faster, but that will make for a weaker mold. If you need only 1 or 2 castings, it's OK to use an accelerator (they can speed up the setting time from 10 hours to 3), but if you want to make dozens of castings, be patient.

With the first thin layer covering the model, blow compressed air over the part (don't get too close) to eliminate any bubbles (Figure G).

When the first layer is just past the tacky stage, brush on another layer of silicone, making sure it's of uniform thickness all over the model (Figure H). For this application, you can add thixotropic agents to increase the silicone's viscosity (but not on the first layer, as these agents make it difficult to get the silicone into all the nooks and crannies for high detail).

7. APPLY SILICONE KEYS.

Once the second layer is applied for the first blanket of silicone, and while that layer is still wet and tacky, start applying the silicone keys. These will help the silicone stay adhered to the inside of the hard plaster mold. While the advantage of this method is that it uses less silicone, the disadvantage is that the thin silicone layer lacks structure and must be married to the plaster mold so it doesn't collapse.

I made these keys from an old silicone mold by cutting small wedges about 1½" long by about a pinky width (make sure that the silicone for the keys and the mold are the same brand — it helps them stick better). For this mold, I set keys in the wet silicone about every 3". As every mold is different, you have to imagine your mold upright and think through the weak spots, where it will buckle, and place the keys accordingly.

Figure I shows the mold with the keys in place — all the silicone is poured (for the first half), and it's

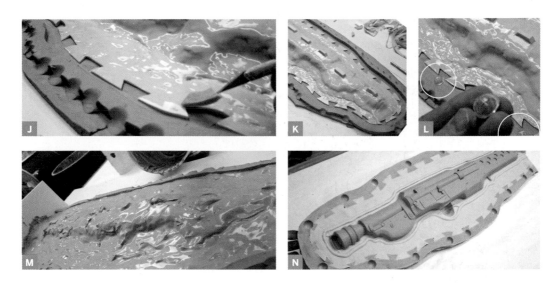

setting up nicely. The silicone doesn't go all the way to the border on the left, but that's OK — next we'll be cutting dovetail keys around the perimeter.

8. CUT THE DOVETAIL KEYS.

After the silicone has cured, trim the edge of the blanket to get a smooth line, and discard the trimmed pieces. Use a sharp X-Acto knife to cut dovetail-shaped keys around the perimeter, gingerly lifting up the edge of the blanket and slicing upward so you don't cut into the clay below (Figure J). This serrated edge will help the silicone blanket register to the mother mold. My blanket here is a wee bit thin at the outer edge. I could probably have trimmed it closer, like about an inch away from the model.

Figure K shows the finished blanket. I've probably used the minimum number of dovetail keys necessary to keep the blanket stable in the mother mold, but you should err on the side of caution and add more than you think the mold might need. Too many keys just makes the mold more stable, but too few and you've wasted a lot of work.

9. ADD HEMISPHERICAL KEYS.

I've left space around the edges of the silicone so I can place hemispherical keys (Figure L). These will register the 2 halves of the mother mold together. I'm using injection-molded ¾" plastic hemispheres, available at any plastics supply store, placed lightly on the clay every 5" or so, just inside the border.

Brush a small amount of vaseline or other mold release onto them to help remove them from the first half of the plaster mother mold.

10. PLASTER OVER THE SILICONE.

After the first blanket of silicone comes the stone-plaster mother mold. The plaster will go on in a couple of layers. The first layer is a thin coat for detail — apply it slowly to avoid creating bubbles. The second is supported by a hemp strengthener.

Figure M shows the first layer of plaster applied over the blanket, the clay, and the keys. It's fairly thick and will take somewhere around an hour to set.

Stone plaster is much stronger than regular plaster. You can get away with using less, which keeps your mold lightweight, but it's still brittle like regular plaster. So the next step is to reinforce it with some spun hemp, available from moldmaking supply stores. The hemp works much like fiberglass, supplying a matrix that increases the plaster's flexibility and makes it shatter-resistant.

Add a layer of hemp, then apply the second and final layer of plaster. Don't wait more than a day between plaster coats, or else the second layer might not stick well to the first. Also note that the first layer of set-up plaster will suck water from the new layer, making it set faster than the first.

Once the second layer's set, turn the whole thing over and gently pull off the clay dam, keeping the model inside the mold (Figure N). Take a moment to study what you've done. Isn't it pretty? The various mold keys are all visible now: the hemispherical keys in the outer ring of plaster, the dovetail keys where the silicone meets the plaster, and the barrier key around the model itself.

11. MAKE THE SECOND SILICONE BLANKET.

Before applying a layer of silicone to the other side of the model, you'll have to clean it and prepare it well — but be careful not to mess with it too much, as you want as tight a registration as possible.

Use a brush and soft damp cloth to remove any clay residue. Apply mold release (or a thin layer of vaseline) to the silicone so that the next layer won't stick to it. Make sure you cover it all, or else you'll ruin your mold. Silicone loves to stick to itself.

Apply 2 layers of silicone exactly as you did in Step 6. As before, use very little accelerator in the first layer, but you can use more, or a higher mix of the kicker, in the second. Remember: the goal is to get a ¼" blanket all around the part.

As with the first blanket, place the silicone key wedges along the center of the model before the second layer of silicone hardens. Cut the dovetail keys from the second layer, gingerly lifting up the edge to avoid cutting into the layer below (Figure O).

Figure P shows the mold with the second silicone blanket done and cut. Now it's time to lay on the plaster for the other half of the mother mold. You're almost there!

12. MAKE THE SECOND HALF OF THE HARD-SHELL MOLD.

To prep for the second plaster, apply mold release to the first plaster half. Again, a thin layer of petroleum jelly works great (Figure Q). Cover the inside of all the hemispherical key indentations because, again, if they don't release, all your work is down the toilet.

Using aluminized tape, available in the plumbing or heating section of any hardware store, build a simple mold dam to contain the wet plaster and give it a nice crisp edge that matches the clay dam from the first blanket (Figure R).

Slowly drip the plaster onto the second blanket of silicone. Again, this first layer is for detail, and the fewer bubbles it has, the better it will hold.

Use your hand to spread the plaster over the

O

P

Q

blanket, making sure it covers everything, especially the silicone wedge keys along the center. But be careful around the edge of the blanket, where the dovetail keys are! This edge may want to lift up, and you don't want to get any plaster under it between the 2 layers of silicone. Better to drip the plaster gingerly around the edge. This is an important point to remember, and the more familiar you are with this whole process, the less likely you are to forget a key step and end up wasting your hard-earned time.

Figure S shows the first layer of plaster. Note how well the aluminized tape dam holds it in. Also note how clean the workspace around the mold is. This type of moldmaking is very detail intensive, and attention to cleanliness during the molding process will quite simply yield a better product.

Allow the second layer of stone plaster (with hemp below it) to dry in all its glory. Once it's dry, all you have to do is pull off the tape and gently pry the halves apart. Since plaster is brittle, care must be taken not to over-torque the mold, lest it crack.

Use 2 screwdrivers or sturdy putty knives, one leapfrogged in front of the other, to proceed down the seam. As you go, listen for the telltale sound of the halves letting go of each other.

Go slowly! You don't want a cracked mother mold before you've even started casting. Making your way down a full side of the mold halves should be sufficient for a proper separation. After a time, you'll hear a sucking sound and see that the 2 halves have popped apart.

The 2 silicone blankets should be somewhat stuck together now, but simply grabbing one of them and pulling it off the other should do the trick (if you've properly applied the mold release). Then pull out the model to reveal your finished mold.

Figure T shows our completed mold. Everything worked perfectly. Note the lack of air bubbles in the positive hemispherical keys on the bottom half. These should register the 2 halves of the mother mold beautifully. And they did. Figure U shows a completed assembly of a resin casting from this mold. To reduce air bubbles on a large part like this,

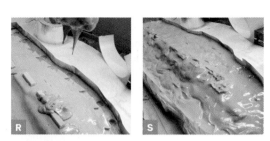

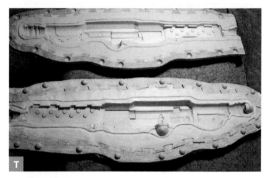

you can first pour some resin in each half and let it cure, then assemble the 2 halves and pour a final resin middle to get the completed casting.

With proper cycling (letting the silicone cool down between castings — heat kills molds), this mold should easily yield 20 or more castings before deteriorating. Because the blanket is thin and the plaster sucks heat out during the resin's curing process, it could even yield 50 castings.

Adam Savage is a lifelong maker, having worked in the theater, fine art sculpture, machine art, robotics, animation, commercials, and films like *A.I.*, *Space Cowboys*, the *Star Wars* prequels, and the *Matrix* sequels. He's taught advanced modelmaking, and has a modelmaking textbook kicking around in his head. For the last eight years he's hosted *MythBusters*. He's the father of 11-year-old twins, and collects and makes movie props and other impossible objects in his spare time.

1+2+3 Sneaky Milk Plastic
By Cy Tymony

Don't have a 3D printer to make plastic parts? Use moo juice instead. Ordinary cow's milk contains a protein called *casein*. When separated from milk by using an acid such as vinegar, casein becomes a moldable plastic material that can be used to create everything from glue to fabric to billiard balls. Make your own custom parts for projects!

1. Cook the milk with vinegar.

Pour 1 cup of milk into the saucepan and warm it to a simmer, not a boil, on the stove.

Next, add 4 teaspoons of vinegar to the milk and stir. After a few minutes you should see white clumps form. When you do, keep stirring a few minutes longer, then turn off the heat to allow the pan to cool.

2. Strain the casein from the milk.

Pour the milk through a strainer into a bowl to separate all the white clumps (this is the casein plastic material), and place them on a sheet of wax paper.

3. Dry and mold the casein plastic.

Dry the casein plastic material by blotting it gently with paper towels until it's dry.

Mold the plastic material into the shape(s) you prefer, and let it dry for at least 2 days. Once it hardens, you can color it with acrylic paint if desired.

TIP: If the casein is too runny to shape in your hands, next time add a pinch of cornstarch to the milk and vinegar mixture. This will make it hold together better.

Going Further

Shape your sneaky milk plastic into a sneaky finger ring that will attract paper U.S. currency! Just hide a small, super-strong (neodymium) magnet inside your ring before it dries and hardens. When the ring is placed close to a legitimate folded bill, the bill will move toward the ring because of the iron particles in the currency's ink.

Excerpted with permission from Sneaky Uses for Everyday Things *by Cy Tymony (Andrews McMeel Publishing, 2003).*

Cy Tymony is the author of the *Sneaky Uses for Everyday Things* book series.

YOU WILL NEED

Metal saucepan
Spoon **for stirring**
Measuring spoons
Strainer
Wax paper
Paper towels

4tsp white vinegar
1c milk

Optional:
Acrylic paint
Cornstarch
Small neodymium magnet

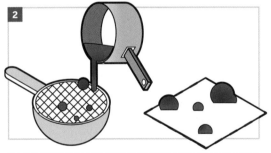

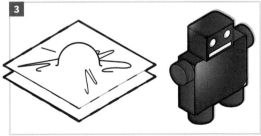

A SIMPLE LIGHT SCULPTURE

Swirling colors for long-exposure photos.
By Jared Bouck

I've always enjoyed long-exposure photography, but sometimes the images can seem predictable. So I decided to build some dynamic light sculptures that would add effects, colors, and patterns to my photos. I put together this simple rig in an afternoon, and a little trial and error taught me how to use it to create a good variety of photographic results.

I used PVC pipe and hobby gearmotors to make this gadget, but the recycling-minded person can build something similar for free using a few deconstructed donor devices and a dose of imagination.

1. Build the frame.

The frame is just a handful of PVC pipe lengths connected with some tee fittings (Figure B, following page). Cut four 24" lengths to make legs, two 4" lengths to connect the tee fittings together, and one 5' length to serve as the upright. Gluing the

frame is optional, as the fittings will hold tight if the pipe is lightly tapped into place.

Once you have the frame, you need to shape a motor mount at the top of the upright. Drill a hole through the pipe 2" below the top using a ¼" drill bit. Then use wire cutters to make 2 shallow, U-shaped snips in the end of the pipe, at the front and back, creating a cradle for the motor (Figure C).

2. Mount the main arm and motor.

Place the medium gearmotor in the cradle, and use zip ties to tightly secure it on top of the upright (Figure D). This is the main drive motor that turns the main arm, which is a balanced length of U channel.

Center the motor sprocket on the outside bottom of the U channel and mark the channel between sprocket teeth on opposite sides. Using a drill bit that matches your screw size, drill 2 holes in the

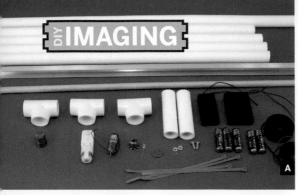

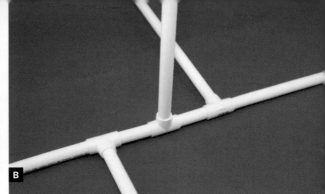

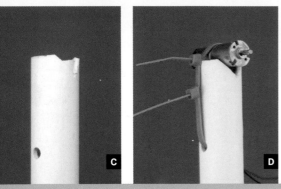

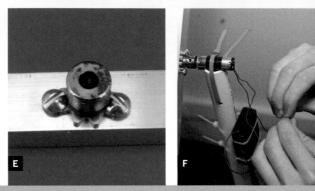

Fig. A: Materials for the light sculpture (but you can easily substitute). Fig. B: The PVC pipe frame holds a 5' PVC pipe upright. Fig. C: Notches in the top of the upright hold the main drive motor.

Fig. D. Zip ties hold the motor in place. Fig. E: Machine screws attach the main motor sprocket to the bottom of the U channel. Fig. F: Zip-tie the main motor battery pack to the upright.

MATERIALS AND TOOLS

PVC pipe, ¾" diameter, 14' length
PVC tee fittings, ¾" (3)
Medium gearmotor I used the Copal 60:1, part #0-COPAL60 from Robot Marketplace (robotmarketplace.com).
Small gearmotor part #GM2 from Solarbotics (solarbotics.com)
¾" aluminum U channel, 55" length from your local home improvement store
½" wood dowel, 36" length
Chain sprocket, #25, 9-tooth, ¼" bore Robot Marketplace #SP-25FB9x.25
Battery packs with switch, 2×AA (1) and 4×AA (1)
LEDs in your choice of colors. For consistency in your photos, I recommend using LEDs that are matched in their brightness. My favorite LED, the RGB color fader, produces really nice effects.
3V lithium watch batteries CR2032 or equivalent, one for each LED
Machine screws and nuts
Mass for counterweights Rolls of pennies work well.
Wire and wire cutters
Drill and drill bits: ¼" and smaller
Hacksaw or PVC pipe cutter
Soldering iron and solder
Electrical tape and zip ties
Camera with long exposure capability
Allen wrench
Optional: PVC glue, ping-pong ball

U channel and attach the sprocket securely on both sides with machine screws and nuts (Figure E).

Attach the completed arm to the main motor shaft, using an Allen wrench to tighten the sprocket's setscrew. Zip-tie the motor's battery pack to the stand (Figure F), and then switch on the motor to ensure that the arm is balanced and doesn't make the stand wobble. Use pennies and electrical tape to balance the arm if needed.

3. Mount the secondary arm and motor.

Measure and mark the center of the wood dowel. Using a drill bit slightly smaller than the drive shaft of the small gearmotor, drill a hole crosswise through the dowel. Fit the motor shaft into the hole, and zip-tie the motor to the main arm just loosely enough to let you slide the motor up and down the arm to produce different effects (Figure G).

The primary arm will now likely be totally out of balance. Attach a counterweight of pennies to the primary arm, opposite the secondary motor, to balance it out again (about 75¢ worked for me).

Zip-tie the small motor's battery pack in the center of the primary arm (Figure H), leaving enough clearance so that the secondary arm won't hit it.

Photography by Ed Troxell

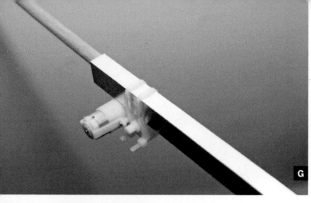

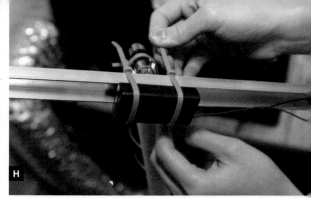

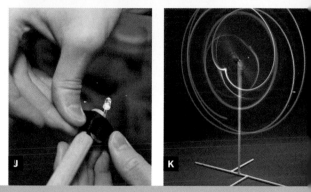

Fig. G: Zip-tie the secondary motor to the main arm with just enough play to let it slide to different positions.
Fig. H: Zip-tie the secondary motor battery pack to the middle of the primary arm.

Fig. I: Run the secondary motor wires through the U channel. Figs. J and K: Tape a glowing LED to the end of the secondary arm, then start experimenting with long-exposure photos.

Run the wires between the secondary motor and battery through the inside of the U channel (Figure I).

Finally, switch on the motor to the secondary arm to make sure it has enough clearance. If you feel satisfied, turn on the motor to the primary arm as well. If need be, make any final balance adjustments to reduce the amount of wobble in the sculpture — though a little wobble won't hurt the overall effect.

4. Light it with LEDs.

Place a 3V coin cell battery between the leads of your LED, accounting for polarity (the positive (+) lead is the longer one). If it lights up, secure it with electrical tape. Then tape the glowing LED and battery to the secondary arm with the LED facing forward (Figure J). You're done!

You can experiment with different colors and placements of LEDs on the secondary arm to produce many different effects. Try putting LEDs on both ends of the arm, or use several LEDs at once.

TIP: LEDs are directional, which can negatively impact the viewing angles for your sculpture. To compensate for this, poke a hole in a ping-pong ball and insert your LED to diffuse the light. Now you can see it from almost any angle.

5. Shoot it!

For clear, sharp shots, always shoot from a tripod. That's the first and foremost rule. To produce a full pattern, take long exposures of approximately 20–30 seconds, depending on the speed of your sculpture (though you can partly compensate for incomplete patterns by placing LEDs on both ends of the secondary arm). To ensure intense, well-balanced colors, use your camera's white balance.

Ultimately the best tip is to take the time to experiment and try several configurations, settings, and exposure lengths to achieve the desired effect. With a little effort, the results can be spectacular.

Additional Experimentation

You can easily reconfigure this project to produce a large variety of patterns and designs by simply changing the orientation of the motors and the placement of LEDs on the arms.

» Find more detailed instructions, alternate configurations, and more projects like this at inventgeek.com/makerart.

Jared Bouck is the CEO of Invent Geek Media and specializes in web marketing and product development.

Photograph by Garry McLeod (K)

DIY IMAGING

PHOTO BOOTH

Make an arcade-style snapshot station for special occasions. By David Cline

If you've ever helped to plan a wedding, you've probably run across ads for traditional photo booth rentals. These booths are overpriced and have limited options for customization, and you only get to play with them for one day! Here's how I took a DIY approach for my sister's wedding.

My photo booth needed to be completely automated, and easy enough for anyone to use — think of all the grandparents! I also wanted a keyboard-only interface, to prevent errors from people clicking on the wrong things.

To meet these needs, I based my booth around a MacBook running Automator, a built-in Mac OS X programming tool that lets you create and run workflows (Figure B), like a drag-and-drop AppleScript Lite. And to make the booth even easier to use, I recorded voice prompts that greet guests and guide them through the workflow. A talking photo booth!

1. Program the Mac.

For the software side of the project, here are the basic steps I programmed into Automator:

a. Greet the user and prompt them to enter their name on the keyboard. Wait for input.

b. Warn the user that 3 photos are about to be taken.

c. Take 3 pictures using the Mac's iSight camera.

d. Assemble an image of all 3 photos lined up in a strip.

e. Print the strip (Figure A).

f. Save the separate images and the photo strip image onto the computer.

g. Loop back to the beginning.

Automator does have its limitations, and there are many details within these steps, so I also wrote some AppleScript routines to make the program fully functional. You can download all the software for a small donation at davidcline.wordpress.com.

Photography by David Cline

Fig. A: Photos print out in traditional photo booth-style vertical strips. Fig. B: Mac OS X Automator controls the photo booth workflow.

MATERIALS

Macintosh computer running OS X 10.5 or 10.6, with a built-in iSight camera or a Mac-compatible webcam (UVC device)
Software: Automator and some AppleScript routines that I wrote You can download all the software for a small donation at davidcline.wordpress.com.
USB keyboard optional, but recommended
Computer speaker
4"×6" photo printer I used an Epson Dash.
4"×6" photo printing paper
Garment rack, about 36"×60"×20" with a frame on top that will carry a dust cover (dust cover not needed)
Shower curtains in bright, solid colors for backdrops
Shower curtain hooks
Stools, about 2' tall (2) not barstool height
Work lights, clip-on (2)
Compact fluorescent (CFL) bulbs (2)
Aluminum foil
Wire, stiff and thin
Computer desk, small
Document bins or cardboard boxes or other ways of arranging the speaker, computer, printer, signage, and printed photos on the desk
Foamcore board or cardboard
Paper cutter
Scissors
Sharp knife

2. Assemble the booth.

To create the booth, I envisioned a simple frame that I could hang backdrops on. I dug through my parents' basement and found an old wardrobe storage rack. It was tall enough to support a backdrop, and the perfect width for 2 people to sit underneath. It also had a metal frame on top for holding a dust cover, from which I could easily hang the backdrop.

For backdrops, I liked the look of Ikea's solid-color shower curtains. They're slightly translucent, so backlighting the material creates a nice glow in the photos.

I hung the curtains on rings from the dust cover support rod along one side. Then I trimmed the bottoms so they didn't touch the ground, and cut them down to half width so there wasn't too much rippling. Because the curtains hooked onto the wardrobe frame so easily, I could offer guests their choice of backdrop colors: red, blue, green, and white.

To enclose the sides of the booth, I cut 1'-wide strips from the extra white shower curtain and hung them on the side of the wardrobe (Figure C, following page).

3. Hook up the printer.

I used an Epson PictureMate Dash, a portable printer that handles 4"×6" paper. Its tiny footprint is great for this setup, and Epson's inks dry instantly and can be handled right away, which is important for a photo booth.

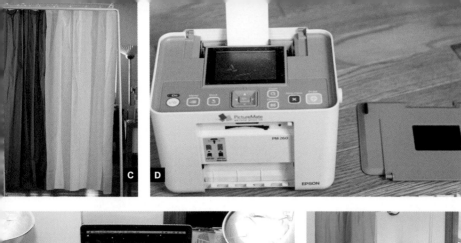

Fig. C: The booth backdrop consists of a tall wardrobe rack and shower curtains. Fig. D: Remove the trays from the photo printer and set the paper guides to 2" wide strips. Fig. E: Cover the CFL bulbs with aluminum foil discs to reduce glare. Fig. F: Raise the laptop so its webcam is at the subject's eye level. Fig. G: The full photo booth setup. Fig. H: Photo strips from happy party guests.

I didn't like the way the printer's paper tray and output tray took up so much room sticking out, so I carefully snapped them off.

I used a paper cutter to cut 4"×6" sheets of photo paper right down the middle, making two 2"×6" strips. The printer's adjustable paper guides accepted a stack of the strips without a problem (Figure D).

4. Rig the lights.

To light the booth, I used 2 clip-on worklights with compact fluorescent bulbs, so they wouldn't get hot. To avoid blinding the guests, I blocked the bulbs with rounds of aluminum foil suspended in front by wires bent around the edges of the lamps' reflectors (Figure E).

You can achieve the same effect with waxed paper or another diffusive material, as long as you've got cool CFL bulbs (not hot incandescents or halogens).

5. Set it all up.

I plugged the printer and speaker into my MacBook and arranged them all using acrylic bins on a computer desk. I clipped one worklight to a bin on each side. Unless the desk is high, you'll need to prop the laptop up, so that it photographs guests head-

on rather than from below, which skews the view (Figure F).

To keep guests' hands off of the laptop, I covered its keyboard and trackpad with a piece of foamcore. I plugged in a USB keyboard at the desk for them to type their names into (Figure G).

That's it! Everything worked as planned for my sister's wedding, and the photo booth was a big hit. In a world of ubiquitous digital photography, there's still something special about posing in front of a curtain and watching your little strip of front-lit photos come out (Figure H).

Set this up at your next party, and you'll be delighted at how much fun your guests will have!

David Cline (davidfcline.com) is a problem-solving, idea-generating, digital-photographing, and video-editing sushi connoisseur who is against functional fixedness. He is currently an information systems major at Drexel University.

ADD VOLUME, JACK

Plug in and turn up any sound-making battery toy. By Peter Edwards

Many cheap, fun sonic and musical toys have built-in speakers and no output jack. This limits their volume, unless you constantly hold them up to a microphone. Here's how to mod these devices so you can plug them in, adjust their volume, and rock out.

The simplest way to install an output jack is to remove the speaker and solder the jack in its place. But every time I do this I regret it, because then the toy no longer works by itself. Inevitably, there will be a time you want to play and there's no amp around.

You can also leave the speaker connected in parallel with a regular audio jack, but then the speaker might act like a microphone and trigger feedback and unwanted noise when you're plugged in.

My favorite solution is to use a switching jack, which automatically disconnects the speaker when you plug in a cord. You can also use a non-switching jack and an on/off toggle that switches the speaker

between standalone and plugged-in modes. But switching jacks are only slightly more expensive than non-switching jacks. Why use two pieces of hardware when one achieves the same effect?

1. Identify your jack's contacts.

First you need to identify three lugs of the jack: tip, sleeve, and switch. There are many different styles of jack available (Figure A, following page), so it's hard to offer general rules for identifying these. Figure B shows our circuit with one common lug configuration. If your jack varies from this one, look for a spec sheet online.

2. Expose the circuit.

Open the device and find its speaker (Figure C). To keep screws in a safe place, you can usually stick them to the speaker magnet.

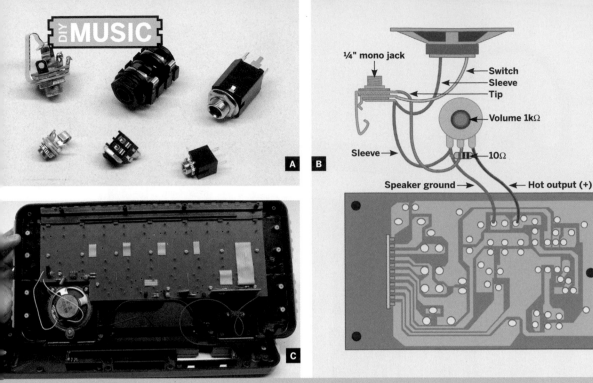

A

B

¼" mono jack

← Switch
← Sleeve
← Tip

← Volume 1kΩ

Sleeve →

─10Ω

Speaker ground →   ← Hot output (+)

C

Fig. A: There are many different styles of ¼" and ⅛" switching audio jacks available. Fig. B: One common connection configuration for a switching audio jack. If your jack varies significantly from this layout, look for a spec sheet online. Fig. C: Open the device and find the speaker. (For safekeeping you can stick the screws to the speaker magnet.)

⚠ **WARNING: Unless you're experienced with electronics, you should only work on circuits that are battery-powered or use a very low-amperage power supply (50mA or so). Poking around inside high-voltage and/or high-current circuits can be fatal.**

MATERIALS AND TOOLS

Audio toy, mono (single speaker) battery-powered or very low amperage

Potentiometer and matching knob A 1kΩ or 10kΩ pot usually works, like item #271-215 from RadioShack ($3, radioshack.com). Some people prefer audio taper (logarithmic) pots, but regular linear pots work fine, too.

Switching audio jack, normally closed (NC) Choose a ¼" jack for guitar cable or ⅛" (3.5mm) for mini/headphone plugs.

Wire, stranded, 22 gauge 3' total is plenty. Scrounge from anywhere, or RadioShack #278-1224 ($7).

10Ω resistor (optional) Attaches across the outer legs of your potentiometer if your output is distorted. RadioShack #271-013 ($1 for 5).

Soldering iron and solder

Drill or Dremel rotary tool

Continuity tester, ohmmeter, or multimeter

Screwdriver, pliers, masking tape, and pencil

3. Identify the speaker ground and hot signal lines.

Two wires connect the speaker to the circuit board: the hot signal that creates the fluctuating sound wave and the stable speaker ground that establishes the signal amplitude. The ground attaches to the power supply (the negative contact in most pro audio gear, but toys can run either way). Once you've identified one wire, you know what the other one is, and you can use some tape to mark them both on your board. Here are clues to check for:

» **Follow the leads to the power supply.**
Follow both speaker leads out to see where they connect, continuing along traces on the board if needed. Whichever wire runs to the power supply is your speaker ground (green wire in Figure D).

» **Use a continuity tester.** Most multimeters have a continuity setting marked with an image of a speaker or sound waves. Disconnect the speaker, touch one meter probe to the positive or negative power supply, and then touch the other probe consecutively to the speaker's contact points on the board. The speaker contact with zero resistance to either power contact is ground.

Fig. D: It's easy to find the speaker ground line in this example (green wire) because it connects directly to the negative battery contact (power). Fig. E: Wire up the potentiometer and jack. Fig. F: If the output is distorted, add a 10Ω resistor across the outer legs of the pot. Fig. G: Mark and drill holes in the housing for the jack and pot. Fig. H: Mount the potentiometer and jack in the housing and close it up. Fig. I: Done!

» **Look for a transistor or audio amplifier IC.** On the board, one of the speaker wires probably connects directly to the output of a transistor or an amp IC such as the LM386. This is your hot signal.

4. Wire up the pot and jack.

On your potentiometer, decide which leg will be the ground and which will be hot (the middle leg is always the sweep). Solder wire connections as follows: jack tip to pot sweep; jack switch to one side of the speaker; jack sleeve to the other side of the speaker and to pot ground; pot ground to speaker ground on the board; pot hot to hot output on the board (Figures B and E). Make the wires long enough to give wiggle room for mounting the controls.

5. Test it!

Plug the device into an amplifier and play it. If it sounds distorted, solder a 10Ω resistor between the hot and ground legs of the potentiometer (Figure F). If the signal is too loud for your liking, you can also add a 1Ω–10Ω resistor between the potentiometer and the output jack.

6. Mount the pot and jack.

Choose and mark locations where the volume and output jack can fit on the toy's casing, then use a drill or Dremel tool to create the appropriate-sized holes (Figure G).

TIP: If you place your holes along a seam, use a high-speed Dremel to remove the plastic with a cutting/routing bit. An ordinary drill would simply push the halves apart or split the plastic.

Mount the hardware (Figure H) and put the knob on the pot. Once that's done, carefully close up the housing, but don't screw it back together yet.

7. Retest and assemble.

In the process of closing the housing, it's likely that you moved around and possibly severed some of your wiring. Test the device again, both plugged-in and unplugged, to make sure it still works. If so, screw the housing back together, and you're ready to rock (Figure I).

Peter Edwards is a circuit-bending and creative-electronics pioneer based in Troy, N.Y. He builds electronic musical instruments for a living at Casper Electronics (casperelectronics.com).

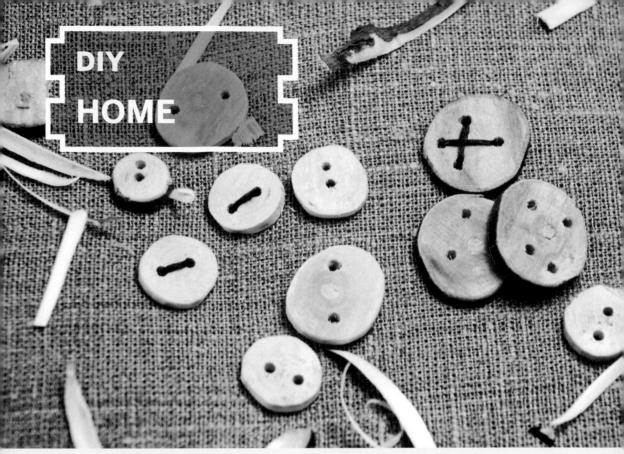

WOODEN BUTTONS

Make your own tree bling from fallen branches. By Kristin Roach

After a big windstorm, it can be a little overwhelming to look at all the branches and sticks that need to be picked up in the yard. Instead of dreading this cleanup task, you can see it as a chance to collect all sorts of great materials for creative projects.

For example: I really like wooden buttons, but the ones from stores are often so highly polished that they barely look like wood at all. Here's how to make nicer ones out of fallen branches that you find yourself. These buttons are smooth, but they retain the beautiful characteristics of the tree they came from — grain, texture, and even bark, if you choose. They are the perfect thing to adorn your totes, shirts, bags, or any other project that needs a little tree bling!

1. Choose a branch.

Find a hardwood branch with a straight section 6"–8" long by ¾"–1½" wide. You can use a fallen

MATERIALS AND TOOLS

**Branch with a straight section 6"–8" long and
¾"–1½" wide, from a hardwood tree Hardwoods
are angiosperms, which produce seeds with some
sort of covering, like fruits, nuts, or acorns. You
want a section with even grain and no forks.
Drill and drill bits: 1/16", 5/64", 3/32"
Sandpaper: grits of 220, 320, 400, (optionally) 600
Knife, small and sharp An X-Acto knife works well, as
does the small spear blade on a pocketknife.
Jigsaw or coping saw and vise or C-clamps
Beeswax (optional)**

branch or cut one yourself, but make sure it has no sticky sap and a small pith (the hollow or spongy center), no more than ⅛" for a 1"-wide branch.

Moisture level is important. Branches that are too dry will snap when bent, and freshly fallen or cut

Photography by Kristin Roach

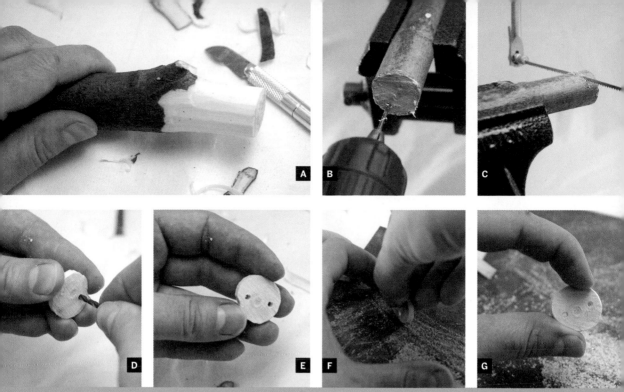

Fig. A: While a standard X-Acto blade will work, I recommend getting a set of carving blades if you plan on doing a lot of button making. Figs. B and C: If you don't have access to a vise, clamp your stick to the end of a table with 2 C-clamps. Figs. D and E: Pre-drilling the holes and then widening them by hand helps prevent splitting. Figs. F and G: Using varied grits of sandpaper will yield a smooth-finished button that retains its rustic look.

branches will need to dry out in the sun for a day or two before being made into buttons. Buttons cut from wood that's too green or wet will crack once they dry all the way. Expect some trial and error with determining the right moisture.

2. Remove the bark (optional).

You can leave the bark on your buttons, but if you want to remove it, use a sharp knife to peel away 2"–3" of bark at one end of the branch, working around it evenly with shallow strokes. It's important that your knife is sharp; if it's dull, removing the bark isn't just a pain — you're also more likely to slip and cut yourself.

3. Drill the holes.

Place the branch in a vise and use the 1⁄16" bit to drill 2 or 4 evenly spaced holes into the cut center of the branch. Keep the drill level with the branch, or your holes will be skewed, and don't drill too close to the perimeter of the branch, or you may break through it.

4. Cut the buttons.

Use a jigsaw or coping saw to cut 1⁄8"-wide circular slices from the branch.

5. Clean out the holes.

Next you'll need to clean the wood pulp out of the holes and widen them. Hold the 5⁄64" bit in your hand and gently turn it through all the buttonholes to widen them. Repeat with the 3⁄32" bit.

TIP: You may want to use a bit of cloth for grasping the drill bit or you could hurt your fingers over the course of making a half-dozen buttons — as I did the first time.

6. Finish the buttons.

Sand the buttons, starting with the 220-grit sandpaper and working through to the finest grit. Depending on how rough you want your buttons, you can do just a few strokes or work all the way up to 600-grit sandpaper to make them really smooth. If you want to finish the buttons with a natural seal, rub on a little beeswax then wipe off the residue with a clean cloth.

Try out a few different ways of sanding and finishing with individual buttons. Once you find a finish you like, go crazy and make a whole set that way!

Making creative projects out of found and on-hand materials is one of Kristin Roach's favorite challenges. You can find more of her crafting adventures on craftleftovers.com.

JAR ORGANIZER

Put those old jars to good use.
By Abe Connally and Josie Moores

Many people have a bunch of jars they can't quite bring themselves to throw away. And why would they? Jars are useful in so many ways. Here's a simple, quick idea for reusing old jars — the Jar Organizer.

Whether you use it for storing screws, seeds, spices, buttons, electronic parts, or anything else, the Jar Organizer is extremely versatile. It's easy to make in an hour, and costs less than $10.

This design accommodates a dozen jars up to 7" tall, but you can use more or fewer jars by changing the length of the 4×4. For stability, the base wants to be at least the length of your tallest jar. Our unit also has space beneath it to store something, so if you don't want anything below yours, you can make it shorter.

We originally used baby food jars, but they ended up self-sealing and we couldn't get them off. So make sure you use jars with proper screw-on lids.

MATERIALS AND TOOLS

Lumber: 2×4, 16" lengths (4) for the base;
    4×4, 18" length to attach the jars to
Jars with screw-on lids an even number, up to 12.
    Plastic is preferable to glass, but not necessary.
Lag screws, ¼"×3" (2)
Wood screws: 3" long (4), ¾" long (24)
Washers, ¼" ID (4)
Paint (optional) leftovers are fine
Drill with drill bits: ¹⁄₁₆", ¹³⁄₆₄", ⁵⁄₁₆"
Wood saw
Tape measure, ruler, and marker

1. Prep your lumber.

Cut four 16" lengths of 2×4 for the base, and an 18" length of 4×4 to attach the jars to. Paint them if you want, and let them dry (Figure A).

Photography by Josie Moores

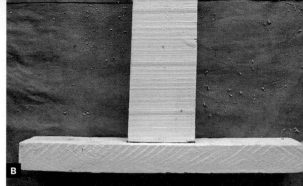

Fig. A: Get all of your parts together: lumber, screws, and jars. Fig. B: Make the "legs" and "feet" of the organizer with 2×4s and 3" wood screws.

Fig. C: Screw the jar lids onto the 4×4. Fig. D: Using 2 diagonal lines, find the center of the 4×4 for the pivot screw.

2. Build the base.

To make the uprights, mark 2 of the 2×4s on their wide faces (4" faces), centered 2" down from the top. Drill a hole straight through each board at your marks, using the $\frac{1}{16}$" drill bit first, and working up to the $\frac{5}{16}$" bit. Make sure your lag screws can pass through the hole. If not, drill it out a little more.

To make the base pieces, measure and mark the centers of the other 2 pieces of 2×4 on their wide faces. Center each upright on top of a base piece, with the hole at the top. Turn the upright so its wide face is parallel to the long side of the base piece, not crosswise (Figure B).

Holding the boards together, turn them upside down like a T. Using the $\frac{1}{16}$" drill bit, pre-drill 2 holes into the joint of the 2 boards. Using the 3" screws, screw the boards together. Repeat this process for the other upright and base piece.

3. Make the jar holder.

Arrange your jars. On each face of the 4×4 you'll have 3 jars, and you want them to be fairly balanced with the jars that will be on the opposite face. Leave enough space between jars to be able to unscrew them comfortably. Mark your arrangement on the 4×4. Remove the lids from the jars and use two

¾" screws to screw each lid into place on the 4×4 (Figure C).

TIP: It's a good idea to pre-drill the lids to avoid cracking them.

At each end of the 4×4, find the center by drawing 2 diagonal lines from corner to corner (Figure D). Pre-drill this center point on each end, using the $\frac{1}{16}$" bit first for an accurate pilot hole, then the $\frac{13}{64}$" bit.

4. Put it all together.

Place a lag screw through its hole in one of the uprights, then through two ¼" washers, then into the 4×4. Screw it in. Repeat for the other side.

Screw the jars into their lids. You're now ready to organize that pile of jumbled-up whatevers into tidy, transparent containers.

📷 **More photos and tips:** velacreations.com/jarorganizer.html

Abe Connally and Josie Moores are a young, adventurous couple living in a secluded off-grid hideaway with their 2-year-old. Their experiments with energy, architecture, and sustainable systems are documented at velacreations.com.

CROCHET CROCKERY

Unique containers made from yarn and resin. By Andrew Lewis

These pretty bowls are perfect containers for sweets, potpourri, or loose change. Amazingly, the main construction material is ordinary knitting yarn, and the construction method is crochet. There are plenty of guides on the internet that can show you how to crochet, but you can always use a store-bought doily if you don't feel up to it.

The real secret to creating these little dishes is water-clear polyester casting resin, which can be purchased from hobby or craft stores. The polyester resin hardens when combined with a catalyst, and fixes the yarn into a rigid shape.

⚠ **WARNING: Work in a warm, well-ventilated area. Polyester resin produces unpleasant fumes, and it won't cure properly if it gets too cold. Wear disposable gloves, goggles, and a face mask.**

MATERIALS

Doily crocheted or store-bought
Water-clear polyester resin and catalyst such as Castin'Craft Clear Polyester (eti-usa.com) or TAP Plastics Clear-Lite (tapplastics.com). **You could also substitute a water-clear epoxy resin and hardener.**
Petroleum jelly aka vaseline
Ceramic or glass dish to use as a mold
Disposable plastic sheet
Disposable plastic cup and bowl
Disposable vinyl gloves
Goggles and face mask

1. Apply a thin layer of petroleum jelly to the outside of the dish that you'll use as the mold (Figure B). This will prevent the resin from sticking to the glass. Imagine you're greasing a cake pan.

Photograph by Sam Murphy; doilies crocheted by Hilary Lewis

Fig. A: Gather your materials and set up in an area with good ventilation. Fig. B: Grease the mold. Fig. C: Mix your resin. Fig. D: Pour the resin mixture into a dipping container. Fig. E: Dip the doily in the resin. Make sure that every bit gets saturated. Squeeze out any excess resin using your hands. Fig. F: Lay the doily carefully over the mold. Fig. G: Sculpt the doily around the bowl paying attention to how you want your edges to look.

2. Place the dish onto the plastic sheet. You may want to apply a little petroleum jelly to the plastic sheet if your doily is bigger than the dish.

3. Measure out your polyester resin into a disposable plastic cup (Figure C). You'll need between 50g–150g (2oz–5oz) of resin for a 6" bowl. The exact amount depends on the absorbency of the yarn you're using. Acrylics generally absorb much less than natural fibers.

4. Add the catalyst to the resin and mix very thoroughly. The ratio of catalyst to resin depends on the manufacturer. The brand I use needs just under 3ml of catalyst per 100g of resin. Note that resin is heavier than water, so 100g is not the same as 100ml.

5. Pour the catalyzed resin into a larger plastic bowl and mix again (Figure D). This ensures that you have a thorough mix, and the larger bowl is more convenient for working with.

6. You don't want to get any petroleum jelly onto the yarn, so change into a fresh pair of disposable gloves.

7. Take your doily and dip it into the bowl of resin. Wash the doily around in the resin, and make sure that every part of the yarn gets soaked.

8. Pull the doily through your closed hand, and wring out any excess resin (Figure E). Repeat this step if necessary. You should be able to open out the doily without any resin dripping off it.

9. Arrange the doily over the mold, and leave it to cure (Figures F and G). Depending on temperature, yarn, and resin mixture, this can take several hours.

10. Once the resin has cured completely, gently ease the crochet crockery from the mold. You might find that a cocktail stick or toothpick is useful for levering stubborn parts free from the mold.

That's it! Once you've made a few bowls you can try your hand at vases and other shapes. Just make sure the shape of the mold permits you to remove your creation once it has hardened.

Andrew Lewis is a keen artificer and computer scientist with special interests in 3D scanning, algorithmics, and open source hardware. upcraft.it

Photography by Ed Troxell

FREEZER BAG HOLDER

Coat hanger keeps zip-lock bags open.

By Larry Cotton

Mama Necessity's offspring, Invention, came to my rescue recently when I was preparing summer veggies for the freezer. I like to use kitchen scissors to slice vegetables, but I wind up with a big pile to pick up, handful by handful, and stuff into freezer bags.

'Twould be nice to skip some handling and let the snipped veggies fall directly into the bags, I thought. But how? Thus from a wire coat hanger was the Freezer Bag Holder (FBH) born. These instructions are for quart-size freezer bags; for larger ones, you'll need more than one hanger and 2 taped joints.

First, clip the hook off the coat hanger and straighten the wire with pliers and your hands. It doesn't have to be perfectly straight, but you need about 38".

Use at least one pair of pliers to make a bend of approximately 60° about ¾" from one end. This will be one part of the overlapping taped joint.

Make the other bends along the coat hanger, working your way down as shown in the illustration: a 6½" foot, a 7¾" riser, a ¾"–1" mount for the binder clip, a 5" cross beam bent out about 2", and a clip mount, riser, and foot symmetrically back down. Don't bend the feet yet, but bend the 2"-deep bow with your fingers, trying to keep the curve nice and smooth.

The clip-mount sections should be just long enough to accommodate the ¾"-wide clips. Make the 2 hairpin bends above these by using your larger pliers to turn the wire 180° around the tip of the smaller pair.

Tweak the bends as you go. When you're done tweaking, the last foot of the hanger should, without much forcing, lie parallel to and overlap the ¾" end you bent first. Cut any excess wire and make a joint by tightly wrapping electrical tape around the overlapped portion.

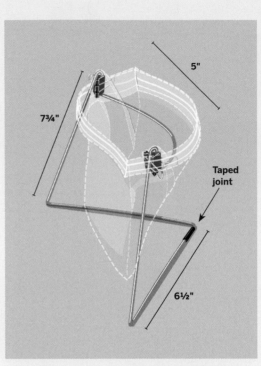

5"

7¾"

Taped joint

6½"

To get the FBH to stand without wobbling, bend a barely noticeable upward curve in the 2 feet, making a base that touches the countertop at 3 points. Finally, put the 2 binder clips over the hairpin bends at the top.

To use the FBH, clamp the freezer bag just below its zip-locking strip, and open the top of the bag as much as possible. Happy bagging!

Larry Cotton is a semi-retired power-tool designer and part-time community college math instructor. He loves music and musical instruments, computers, birds, electronics, furniture design, and his wife — not necessarily in that order.

Illustration by Damien Scogin

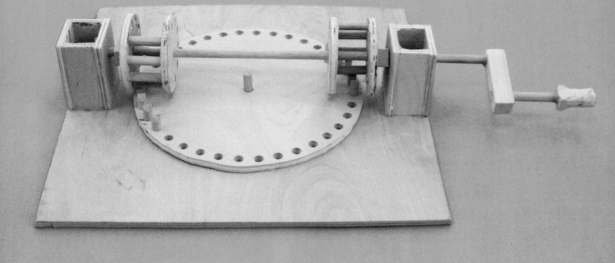

DA VINCI RECIPROCATING MECHANISM

Re-create Renaissance tech to convert rotation into a square wave. By Alan Federman

Last year I was blown away by the clever mechanisms displayed in the Leonardo da Vinci exhibit at the Tech Museum of Innovation (thetech.org) in San Jose. I was especially taken by da Vinci's simple mechanism for powering a sawmill with a water wheel. I made my own tabletop model of the mechanism, and it never fails to gather a crowd when I show it off.

The typical mechanism for converting between rotation and reciprocal motion is a flywheel and a crankshaft, like on a steam locomotive. But da Vinci's simple device starts with 2 modified cage gears (aka lantern gears) that rotate on a common shaft. Each cage has half of its teeth missing, on opposite sides of the shaft from the other gear, so

that when turned, they alternately engage with the pegs on opposite sides of a large wheel.

When you turn the shaft at a constant rate, this mechanism generates square-wave motion, rather than the sinusoidal motion of a piston. This is because each cage gear turns the wheel at a constant speed, and it lets go right before the other one comes pushing full-speed to turn the wheel the other direction.

Cut the Pieces

I built my initial prototype by hand, and it took a lot of trial and error to get the peg spacing right. So I made a more precise second version by laying out the plywood pieces in Google SketchUp

Photography by Ed Troxell

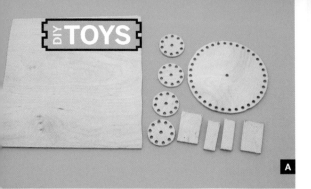

A

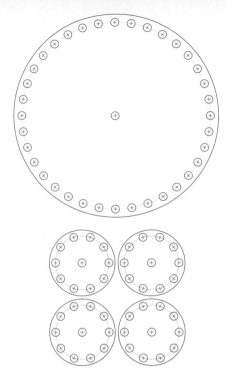

B C

Fig. A: The main plywood pieces for the base, large gear, 2 cage gears (aka lantern gears), and one of the 2 towers. Fig. B: Dowels, wooden washers, and crank assembly parts. The same dowel is used to make the crankshaft and gear pegs. Fig. C: Gear layout drawing. Drill the holes before you cut out the gears; you really only need 10 holes in the large wheel, and 4 in each cage gear disc.

MATERIALS

¼" birch plywood, 12"×18"
¼" wood dowel, 36" long
½"×½" square wood molding, at least 4" long
    aka square dowel
Wood glue

TOOLS

Band saw or jigsaw
Drill press with circle cutter and ¼", 17/64",
    and 9/32" bits
Laser cutter (optional) **instead of the band/jig saw
    and drill press**
Small saw for cutting dowels
File, sanding block with medium-grit sandpaper,
    or Dremel with sanding bit
Glue gun with hot glue
C-clamps (2)
Spray adhesive and computer with printer **for the
    paper templates, if you don't use a laser cutter**

(sketchup.google.com), then importing them into CorelDraw and using the Corel files to cut them on an Epilog laser cutter.

The main plywood pieces you need to cut are 4 identical discs for the cage gears and a larger disc for the wheel (Figure A). Each cage gear disc is 2.06" in diameter, with 10 equidistant ¼" holes spaced 0.84" away from its center point. The wheel is 6.4" wide with 36 equidistant ¼" holes spaced 2.95" from its center.

To match da Vinci's plan, I also cut 8 interlocking rectangular pieces that assemble into the 2 towers that hold the crankshaft. You can download my SketchUp and Corel files at makezine.com/24/davinci, along with full-size PDF templates that you can use to make the pieces by hand (Figure C).

If you're cutting by hand, print out the 3 full-size templates, cut the shapes out of the paper, and temporarily affix them to the plywood with spray adhesive. Drill all the holes before you cut the discs out of the plywood, 17/64" holes in the center and ¼" holes around the perimeter. Functionally, the mechanism needs only 26 holes cut, 4 in each cage gear disc and 10 in the large wheel. But I cut holes all the way around the perimeters of all the discs, 76 total, so that I could use the parts for other purposes, such as a crank assembly for a crane.

In addition to the discs, you need plywood pieces for the base and the towers. The base is simply a 10" square with a 17/64" hole in the center to hold a vertical axle peg for the wheel.

The 2 towers sit on either side of the wheel and

Diagrams by Alan Federman

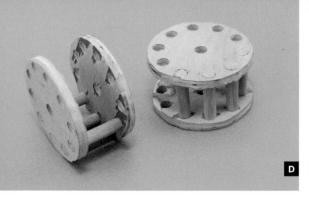

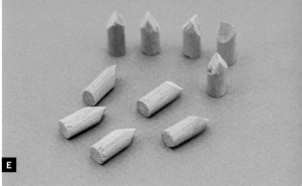

D E

F G

Fig. D: The 2 lantern gears after assembly.
Fig. E: It's helpful to taper the pegs for the large gear;
this wil help them to mesh smoothly with the cage
gears. Fig. F: The main gear assembled.

Fig. G: To measure the proper height for drilling the
mounting towers, mesh a cage gear with the large gear's
pegs, and mark where the shaft will meet the tower.
Then drill slightly above your mark.

hold the drive shaft above it horizontally. I used
the laser cutter to fashion them out of interlocking
pieces of ¼" plywood, and you can also make them
by cutting and gluing simple plywood rectangles;
the PDF templates include plans for both. You can
also cut the towers out of solid blocks. But don't drill
the tower holes yet; you'll do this later, to ensure
they're at the correct height.

From the ¼" dowel, you need to cut one 12" length
for the drive shaft, one 2" length for the hand crank,
eight 1¼" pegs for the cage gears, one 1" peg for the
wheel's axle, and ten ⅝" pegs for the wheel's gearing
(Figure B). If you mark the pegs for cutting all at
once, include some extra length for the width of the
saw blade, or else the pegs might wind up too short.

From the ½" square dowel, cut and drill 2 thin
slices to make washers that will fit over the drive
shaft. Also cut a 2" block, and drill 2 parallel ¹⁷⁄₆₄"
holes through the block at opposite ends. Then cut
another short piece, file it to make a round knob,
and drill a blind ¹⁷⁄₆₄" hole halfway into the knob.

Assemble and Adjust

Build the 2 lantern gears by gluing four 1¼" pegs,
positioned contiguously, between each pair of the
smaller discs. Use wood glue and make sure the

discs are aligned parallel (Figure D).

For the wheel axle, glue a 1" peg into the hole
in the center of the base. Don't glue the towers
together yet.

Sand or file one end of each ⅝" peg so that it's
slightly rounded and tapered on 2 opposite edges;
this helps the pegs engage with the cage gears
(Figure E). Glue the pegs into the wheel, 5 in a row
on opposite sides, with 13 empty holes (or undrilled
hole positions) between them in each direction
(Figure F). The pegs should sit flush with the under-
side of the wheel, stick up ⅜" from the top face,
and be oriented with their tapered edges facing
the neighboring pegs. Let everything dry overnight.
Get some sleep; tomorrow we have fun.

Fitting the gears together is a bit tricky and takes
some patience. First, fit the wheel over the axle.
Then mesh a lantern gear with the wheel's pegs on
one side and use it to find the correct height for the
crankshaft. This is the hardest part, but it's critical
to smooth operation.

The best way is to drill 2 of the wide tower pieces
a little above where you think the crankshaft should
go. You can mark the point by running a ballpoint
pen filler through the cage gear (Figure G). Use a
⁹⁄₃₂" bit so that the crankshaft spins freely.

Fig. H: It's important to drill the tower holes squarely, so the crankshaft won't act squirrely. Fig. I: Mounting the first tower. If it's a bit too high, you can sand down the bottom. If it's too low, you can shim it up.

Fig. J: Adjusting the height of the peg teeth and rounding them slightly with a file is an alternative way to fine-tune the operation. Fig. K: Once everything works properly, add the crank.

Then repeatedly clamp the 2 tower pieces down, test the drive shaft through them, and sand down the tower's bottoms evenly until you reach a good height. (As an alternative, you could make the tower heights adjustable with a tongue-and-groove and a small setscrew, but this cheats the nice all-wood feel of the project.)

Once you have the proper height, stack the other tower pieces and cut and drill them to match (Figure H). Glue the tower pieces together.

For final adjusting, hot-glue the towers to the base (Figure I). Assemble the crankshaft through the towers, hot-gluing the cage gears in place. Work the mechanism in both directions and watch closely to see where any pegs stick, then sand down the pegs and cage gear dowels as needed (Figure J).

When everything works smoothly, mark the precise locations of the cage gears on the crankshaft, the towers on the base, and the shaft where it exits the towers; you'll glue the wooden washers here to prevent the shaft from sliding back and forth. You can put the washers inboard or outboard of the towers; outboard makes it easier to test-glue and adjust the washers.

Using wood glue, permanently reglue the towers to the base (rethreading the crankshaft if necessary

to add your washers), and reglue the cage gears in place. Hot-glue the washers in place, test the mechanism again, then reglue them permanently.

Finally, glue the crank pieces onto the longer protruding end of the crankshaft, and crank away (Figure K).

➕ Download the Sketchup and Corel files as well as a full-size template from: makezine.com/24/davinci

---

Alan Federman lives in San Jose, Calif., and is an instructor at the TechShop in Menlo Park (techshop.ws).

Picture Frame Messages
Save PowerPoint slides as JPEGs and you can load any phrase you want into a digital picture frame. First measure the height and width of your frame's screen, double them, and enter the doubled dimensions in Page Setup — this gives a higher-resolution image. Then type your slides, export them by selecting Save As → *.jpg, and copy the JPEGs to your picture frame. You can even upload my alignment slide at makezine.com/24/tips to find the true edges of your frame, and then adjust your slides to fit your frame perfectly. —*Herschel Knapp*

Find more tools-n-tips at makezine.com/tnt.

BUCKET FULL O' POCKETS

Make your tool bucket even handier with a nifty cover. By Abe Connally and Josie Moores

For almost any project, organization goes a long way toward making things go more smoothly. A tool bucket helps with that organization. It's portable, and it keeps your tools organized, visible, and accessible.

This caddy cover, which fits over a 5-gallon bucket, costs about $10 and will take about a day to make. If you can get your hands on a heavy-duty sewing machine, we strongly recommend you do so — it will save you considerable time.

1. Customize your design.

When making your cover, it helps to have an idea of what kind of tools you want to store in your bucket. Make a mental list so you can customize the design to meet your needs. Adjust the pocket sizes according to what you'll store in them. For heavy, large objects you'll want to make the pockets as tall as possible, and you'll need to increase their depth.

MATERIALS

5-gallon bucket
Heavy fabric, 3'×5' **such as canvas or denim**
Elastic, 1" or 2" wide, 40" in length
8" straps, 1"–2" wide (3) **Nylon straps work well, or you can use excess fabric, hemmed to prevent fraying.**
Snap buttons (3)
Thread

TOOLS

Scissors, sharp
Needle
Pins
Tape measure
Sewing machine **preferably heavy-duty**
Pen or fabric chalk

A **B**

C **D** **E**

Fig. A: The basic items necessary to create your caddy cover, including 2 sizes of rolled white elastic. (Not pictured: 5-gallon bucket, 8" straps, and sewing machine.) Fig. B: Pin the 23"×22" panels to prepare for hemming. Fig. C: After pinning, line up the edges of the pocket piece with the sides of the panel. Fig. D: The elastic band will let you hang tools for easy accessibility. Fig. E: The finished panels pinned and ready to be stitched together.

2. Hem your panels and pockets.

When we say "hem" we mean fold over the edge of the fabric 1" and sew the folded piece to the main piece of fabric. This helps prevent fraying and makes the edges a little stiffer.

It helps to first pin down where you'll be sewing. This keeps the fabric in place as you sew and helps you visualize where you'll be sewing (Figure B).

2a. Cut 2 pieces of fabric, each 23"×22". Hem the 22" sides of both pieces. These are the tops and bottoms of your panels. Hem the 23" sides of both panels, leaving 9½" at the top unstitched (these sections will be hemmed at the end of the project).

2b. Cut pieces to the following 3 sizes and hem on all sides: 6½"×28" (the final hemmed piece will be 4½"×26"); 8"×19" (final piece will be 6"×17"); and 5½"×13" (final piece will be 3½"×11").

3. Sew the pockets.

Figure C shows a hemmed pocket piece pinned and ready to be sewn to the first panel.

First Panel

3a. Lay one of the large panels on a flat surface,

hemmed edges facing down, with the unstitched sections at the top. Use a pen or fabric chalk to mark the bottom edge from left to right at 2", 4", 7", 13", and 17" (there should be 3" remaining at the right edge). Place the 4½"×26" pocket piece on a flat surface, hemmed edges down. From the left, mark the bottom edge at 3", 6", 10", 17", and 22" (there should be 4" remaining at the right edge).

3b. Place the 4½" pocket piece horizontally on top of the panel, lining up the left and bottom edges of both pieces, and pin the 2 pieces together along the left edge, keeping the pin vertical (parallel to the left edge). Line up the first mark on the large panel with the first mark of the 4½" pocket piece and pin it vertically. Do the same for the next 4 marks, then pin the right edges of the pieces together.

3c. Sew vertically along the 7 pinned lines, then sew the 2 pieces together horizontally along the bottom edge. Because of the depth of the pockets, there will be excess fabric. Try to bunch this up evenly along the whole pocket or into the corners of each pocket.

Second Panel

3d. Take the other large panel and lay it on a flat

Photography by Josie Moores

surface, hemmed edges down, with the unstitched sections at the top. From the left, mark the bottom edge at 6", 8", 11½", 13", and 17" (there should be 3" remaining at the right edge). Take the 6"×17" pocket piece and place it on a flat surface. From the left, mark the bottom edge at 9" and 12½" (there should be 4½" remaining on the right side).

3e. Place the 6" pocket on top of the panel, lining up the left and bottom edges, and pin the 2 together along the left edge. Match up the first mark on the large panel with the first mark of the 6" pocket and pin it vertically. Line up and pin the second mark, then line the right edge of the 6" pocket with the 11½" mark on the panel and pin.

3f. Sew along the 4 pinned lines, then sew the 2 pieces together along the bottom edge, trying to bunch the excess fabric evenly or into the corners of the pockets.

3g. Place the 3½"×11" pocket piece on a flat surface and mark the bottom edge 7" from the left (there should be 4" remaining to the right). Place the 3½" pocket on top of the panel, with bottom edges aligned and the pocket's left edge lined up with the 13" mark on the panel (1½" to the right of where the 6" pocket ends) and pin it vertically to the panel. Line up and pin the mark on the 3½" pocket with the last mark on the panel (at 17"), then line up and pin the right edges of the panel and the 3½" pocket.

3h. Sew along the 3 pinned lines, then sew the bottom edges together as you did with the previous pockets.

4. Sew the elastic.
4a. Cut 2 pieces of elastic 20" in length (you can use 1" or 2" elastic, or one of each). Line up one piece of elastic horizontally on the pocket side of one of the panels, 3½" from the top (Figure D).

4b. Mark lines vertically on the elastic. The distance between the marks depends on what you'll want to hang. Mark lines ¾" apart for screwdrivers, 1½" apart for pliers, 3" apart for scissors, and so on.

4c. Sew along these lines, making 2–3 passes for extra strength.
  Then follow the same steps for the other piece of elastic on the other panel.

5. Add the buttons and straps.
Three 8" pieces of nylon strap (or hemmed pieces of excess caddy fabric) are used to attach rolls of tape (such as electrical, duct, or teflon), and small objects with an opening that a strap can fit through.

5a. Sew one part of a snap button to one end of each strap, centered about ¾" from the end. With the button end below, place each strap vertically on the panel (above the 3½" or 4½" pockets), spaced a few inches apart and 10½" down from the top.

5b. Make sure that the snap button piece on the bottom of each strap is facing outward. Sew the other part of the snap button to the top of each strap, securely sewing through the panel as well.

6. Sew the panels together.
Figure E shows the panels placed together with all parts attached, pinned and ready to be sewn.

6a. Using sharp scissors, carefully cut a horizontal slit on both sides of each panel, 1" long, where the stitched and unstitched parts of the sides meet, so that the unstitched sections are no longer folded in. This will give you a total of 4 slits. Place one panel on top of the other, so that the pockets and elastic are facing in toward each other.

6b. Pin the 2 left edges of the unstitched sides together, then do the same for the right edges. Sew the 2 pinned sections together, being sure to sew along the 1" slits you just made to prevent fraying.

6c. Turn the caddy inside out, so that the pockets are on the outside. It should resemble a tube.

7. Put the caddy on the bucket.
7a. Line up the 2 joins of the panels where the handle comes out of the bucket. Place the caddy over the bucket, and feed one panel through the handle.

7b. Fold the elastic sides down into the bucket, and place your tools in their new bucket home. Your tool caddy is complete and rarin' to go!

HOMEMADE SEED STARTERS

Avoid transplant mortality by making soil blocks. By Abe Connally and Josie Moores

Gardening can be fun and rewarding, but some aspects can really hurt your confidence. Transplant mortality is one of those inevitable discouragements. It really pays to evaluate your transplant system before you kill too many of your lovely little garden starts.

One of the best systems we've seen is soil blocks. They have many advantages, but the main one is that they greatly reduce transplant shock, leaving you with more survivors in your garden. They won't create root balls like starter pots or trays, they don't limit roots in the soil like peat or paper pots, and they don't destroy roots the way flats do.

Making soil blocks requires a soil block maker, or *blocker*. The blocker compacts soil into plugs that come with a preformed depression for placing your seeds. Getting your plants off to a good start has never been easier!

The soil blocker is easy to make in about a half-hour. With practice, you'll be able to make soil blocks at a rate of 3 or 4 per minute. The cost should be less than $5, or free if you have various materials lying around.

MATERIALS AND TOOLS

Smooth can, metal or plastic, with lid 2"–4" in diameter, depending on the plants you're starting
⅜" eye bolt, 6"–8" long
⅜" nuts (2 or 3) for eye bolt
Washers (2) for eye bolt
Several large trays
    We use plastic bakery containers.
Soil mix
Drill and drill bits: ⅛", ⅜"
Hacksaw
Pliers
Wrenches (2)

Photography by Josie Moores

Make Your Soil Blocker

Almost every part of your blocker can be salvaged. Paint cans, medicine bottles, or just about any smooth container will work for the cylinder. The eye bolt serves as a plunger handle, and you could just as easily use a long bolt with a wooden handle. The bakery containers are convenient trays, as they come with clear lids and are reusable.

1. Make the cylinder.

You can make soil blockers any size you want, depending on the seeds you have. A 2" block (½-pint can) is best for most seeds. We like the 4" size of the quart can for large seeds like squash and beans. Smaller blocks are good for starting faster-growing, smaller seeds like lettuce, greens, and onions. If you go with a smaller can, you may want to reduce the size of the eye bolt as well, cutting it down to 3"–6" long.

The main thing with the can selection is to avoid a corrugated can, like a soup can, as the soil won't release easily. Before starting, be sure to clean the can well.

Using the hacksaw, cut off the bottom of your can, about ⅛" to ¼" from the bottom. Cut slowly, and make sure you get a good, straight cut. The can bottom will become the plate that presses the soil blocks.

2. Make the press plate.

With your pliers, go along the edge of the cut-off can bottom and bend the ⅛"–¼" remaining part of the can toward the center. This should give you a nice, smooth circular plate without any sharp edges.

Mark the center of your plate, then drill a hole at your center mark, starting with the ⅛" drill bit for a pilot hole, and then stepping up to your ⅜" bit.

Drill a hole in the center of the lid, and trim it to fit within the cylinder if necessary.

3. Assemble the plunger.

Put one nut on your eye bolt, and thread it about 1½" to 2" from the end of the bolt. Place a washer on the eye bolt, followed by the lid of the can. If the lid has a protruding rim, make sure the rim faces the eye of the bolt.

Now, slide on your press plate and follow it with a washer. Thread your last nut onto the eye bolt. The very last nut should be flush with the tip of the bolt. Tighten the first nut back against the last nut

you put on, sandwiching the other parts tightly between (Figure A, following page). You should now have a very stable plunger (Figure B). This plunger slides inside the main cylinder of the can.

NOTE: If you're making a bigger blocker, put 2 nuts on the tip to make a deeper depression for larger seeds.

Press Your Own Soil Blocks

4. Mix the soil.

This recipe (below) is almost foolproof. One "unit" can be any sort of can or bucket, depending on the amount of soil you need (start small). There are lots of different recipes online, so feel free to experiment. You can also buy commercial mixes, but we haven't tried those.

Sift all ingredients before mixing. Mix the peat with the lime or wood ash first. Mix the sand or perlite with the fertilizer. Then mix everything together.

ELIOT COLEMAN'S ORGANIC SOIL RECIPE

30 units peat
⅛ unit lime or ½ unit wood ashes
20 units coarse sand or perlite
¾ unit organic fertilizer (equal parts blood meal, colloidal phosphate, and greensand)
10 units good garden soil
20 units well-aged compost

---

5. Set up a nice large work area.

You'll want a hard surface like a concrete slab to dump your soil mix on, and then another area to keep your trays and blocks once they're made. Dump your mix onto your hard surface. Patios are good for this, as well as outdoor worktables.

6. Wet your mix.

Be careful with this step, as most people won't add enough water initially. You want your mix thoroughly moist, almost dripping water. If your materials are fairly dry, a good ratio is 3 parts mix to 1 part water by volume.

Make a large pile of your wetted mix. The pile should be an inch taller than your blocker.

7. Press a soil block.

Place the plunger into the cylinder of your blocker. The eye should come through the top toward you.

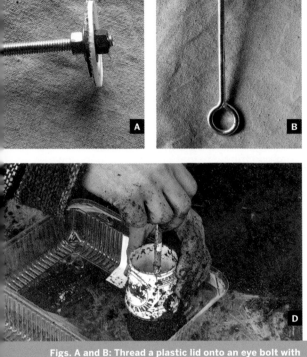

Figs. A and B: Thread a plastic lid onto an eye bolt with 2 nuts and a big washer, and you've got your plunger. Fig. C: Ram the soil blocker into a special soil mix, then pack firmly with the plunger to form a soil block.

Fig. D: Gently eject the soil block. Fig. E: Space your soil blocks ¼" apart, then plant and water. The bolt and nut make a nice divot for planting; clear, lidded bakery trays admit sunlight and retain moisture.

Now dip the blocker in water.

Next, ram the blocker bottom-first into the mix until it hits the hard surface below. Give it a few twists to make sure it's making contact with the surface.

Pack the soil lightly but firmly with the plunger (Figure C). This should "seat" the soil in the blocker.

Gently lift the blocker (with soil) out of the pile of mix. The block of soil should stay inside. If it falls out, try again and gently twist and tilt the blocker as you lift it. If the soil won't stay in the blocker, add a bit more water and pack the soil a bit harder.

8. Eject the soil block.

Place the blocker with the packed soil facing down in your tray. With one hand, press down with the eye bolt; with the other hand, gently raise the cylinder, twisting it as it rises (Figure D).

You should be left with a nice, compacted soil block on your tray. The eye bolt and nuts will leave a good depression to place seeds in.

Dip your blocker in water after each block to help keep it clean and ensure a good release. Space each block about ¼" from the others (Figure E).

9. Plant and water your seeds.

Once you've made several blocks, place a seed or 2 in each block depression, then cover the seeds lightly with a bit of soil mix. Gently water the blocks with a very fine mist. Put a bit of water in the tray, ¼"–½" deep, to help keep the blocks moist.

Cover your blocks with a clear lid if your tray has one; if not, plastic wrap will do just fine. Be sure to keep your blocks moist, watering them once or twice daily. Once the seeds sprout, you can remove the lid.

Depending on your plants, you can usually transplant them 2–4 weeks after they sprout. Handle the blocks with care during transplanting, and use a spatula to make it easy to lift the blocks out of the tray and into their new homes.

Resources
» *The New Organic Grower* by Eliot Coleman
» More on soil blocks:
  • pottingblocks.com
  • toppertwo.tripod.com/soil_blocks.htm
  • velacreations.com/soilblocks.html

GREENHOUSE CONTROLLER

Build a temperature-switched power outlet and save some green. By Andrew Lewis

Photography by Ed Troxell

My dad is a keen gardener. When he needed a new temperature-controlled switch for his greenhouse heater, I knew it was my chance to offset my reputation as the Destroyer of Plants.

I had seen thermostatic plugs online for $40, but I was sure I could come up with something better. Here's how I built a thermostatic controller for my dad's greenhouse for about $25, with some scrounging.

I built the controller around an inexpensive thermostatic control board, which I boxed up with a grounded power outlet for it to control, and a wall wart adapter to supply it with DC power. Plug the box in, and the board switches the power outlet on and off based on ambient temperature. The ground prong on the device's plug simply connects through to the ground receptacle on its controlled outlet. The plug's neutral conductor splits between the wall wart's neutral blade and the outlet's neutral slot, and its hot conductor splits between the wall wart's hot blade and the outlet's hot slot, connecting indirectly to the outlet through the thermostat board's common (COM) and normally open (NO) terminals (Figure A, next page).

This arrangement supplies constant power to the board as it switches the hot connection to the outlet on and off. I also used an audio jack and plug to let me connect the thermistor (temperature sensor) outside the box and position it in different locations.

1. Prepare the power adapter and thermostat.

Depending on what sort of wall wart you have, either unscrew its plastic housing, or just break it away using brute force. Cut its DC output power cord to about 12".

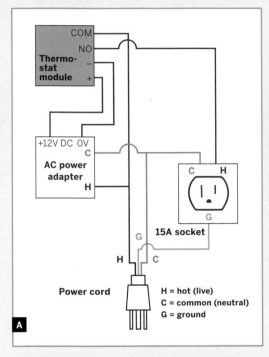

COM
NO
−
+

Thermo-stat module

+12V DC 0V
C

AC power adapter

H

C H

G

15A socket

G
H C

Power cord

H = hot (live)
C = common (neutral)
G = ground

A

MATERIALS AND TOOLS

Thermostat kit, Velleman item #MK138 from Abra Electronics (abra-electronics.com**), $10**
AC power adapter ("wall wart"), 12V DC Countless small appliances use these.
Grounded power outlet with leads, 125V AC, 15A, panel-mount such as Leviton #1374, available at electrical supply or hardware stores, $1–$2
Plastic electronics enclosure, approx. 7"×5"×3" such as #270-1807 from RadioShack (radioshack.com**), $6**
Mono audio plug and jack, matching sizes, panel mount such as #274-319 and #274-346 from RadioShack ($3 and $4 each for a 4-pack)
LED holder #LMH-1 from Abra Electronics (keyword search: "LED holder 5mm"), $0.11
Grounded power cord, 6' without plugs
Insulated wire, 22 to 24 gauge, 4' is plenty
Heat-shrink tubing, various sizes #278-1610 from RadioShack, $4
Grounded power plug, 15A
Fuse, 3A, and in-line fuse holder #270-1009 and #270-1238 from RadioShack, $2 and $3
Cable ties
Blue LED, 5mm (optional) #LED-5B from Abra Electronics (keyword: "5mm blue LED"), $2
Soldering iron and solder, hot glue gun and glue, files, screwdriver, wire cutters and strippers, drill and drill bits, rotary tool with cutting bit, multimeter

Follow the Velleman kit instructions to assemble the thermostat board with one modification: instead of soldering the LED directly to the board, attach two 18" lengths of insulated wire so that the LED can be mounted remotely later (Figure B).

2. Prepare the project box.
The front panel of your project box will house the power outlet, audio jack, LED (in its holder), and the thermostat's potentiometer knob. Position these according to your personal preference, making sure you can glue the thermostat board inside the box so that its knob will extend out where you want it. Drill holes in the front panel for the jack and LED holder and for the knob to fit through, and use a rotary tool to cut a hole for the power outlet.

In the rear panel, drill a hole just large enough for the power cord to slip through snugly.

3. Wire in the electronics.
Cut a 12" length of power cord and cut away the outer sheath, exposing the black (or brown), white, and green wires. These are the cord's hot, neutral, and ground wires, respectively. Save them for later. Cut away about 8" of the outer sheath of the remaining cord, and strip the ends of each wire to expose about ½" of copper.

For the rest of the wiring, you'll be working in and around the box, but don't glue the board and adapter inside yet. Connect the hot (black or brown) lead from the power outlet to the COM screw terminal on the thermostat board (Figure C). Fit the power cord through the back of the box, and solder the ground (green) wire from the socket to the ground wire of the cord, insulating the connection with heat-shrink tubing (Figure D).

Solder the hot wire from the power cord to one blade of the power adapter plug (Figure E). Solder one end of the black wire you cut to the same blade, and insulate the connections with heat-shrink. Connect the other end of the black wire to the thermostat's NO terminal.

Solder the neutral lead from the power outlet to the other blade of the power adapter plug, solder the white wire from the power cord to the same blade, and insulate the connections with heat-shrink (Figure F).

Separate and strip the DC output wires from the adapter. Use a multimeter to determine which is positive and which is negative, and connect

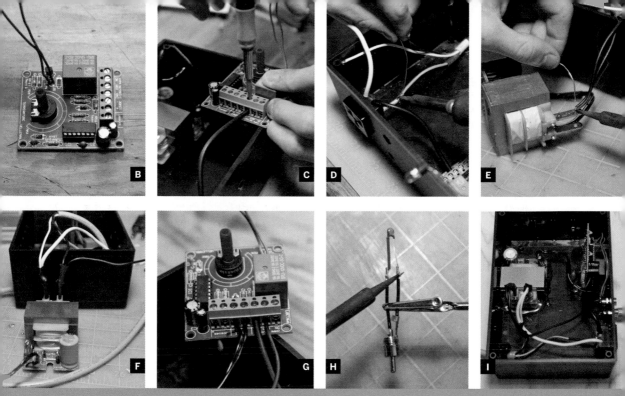

Fig. B: Assemble kit with extension wires for indicator LED. Fig. C: Connect hot power lead to COM terminal on thermostat board. Fig. D: Solder power cord ground lead to controlled outlet ground. Fig. E: Solder hot power wire to wall-wart prong. Fig. F: Insulate wall-wart plug connections with heat-shrink. Fig. G: Connect DC output to board. Fig. H: Solder thermistor to audio plug. Fig. I: Glue components inside box.

them accordingly to the (+) and (−) 12V DC screw terminals of the thermostat board (Figure G). Be careful! Wiring the board in reverse can damage it.

Connect the LED to the wire stalk you soldered to the board earlier, and plug it into the LED holder in the front panel. I substituted a blue LED to look cooler than the red one from the kit. Solder short wires to the audio jack terminals in the panel and connect them to the sensor terminals on the board.

Solder the small blue thermistor from the Velleman kit to the phono plug (Figure H), and insulate the connections with heat-shrink. You can attach the thermistor directly so it will sense the temperature near the front panel, or use longer wires to let you position it some distance away. Insert the plug into the jack on the panel.

At the free end of the power cord, solder an in-line fuse holder with a 3A fuse to the hot wire. Insulate the connections with heat-shrink, but leave the holder itself exposed so you can change the fuse. Connect the 3 wires to a grounded power plug.

4. Hot-glue and test.

Hot-glue the adapter and thermostat board into the box, with the adapter toward the back and the board positioned so the potentiometer pokes through its hole in front (Figure I).

Use cable ties and hot glue to fix the power cord in place, so that it can't be accidentally pulled away. Use more cable ties to tidy up the wiring, and more hot glue to steady the front and rear panels. Put the lid on the box, and get ready for testing.

You can test the unit by plugging in an electric light or other small appliance. The LED should activate when the temperature falls below the level set by the potentiometer. Turn the pot down until the LED comes on, then warm the thermistor in your hand. If everything is working, the LED should go off when the sensor registers your body heat.

5. Add the finishing touches.

The thermistor control ranges from about 41°F–104°F, so you can mark a dial on the front of the panel if you want to make things look neat.

To control heavy electrical loads, replace the relay in the thermostat kit with a suitable alternative. Most greenhouse heaters are low-power (~300W), so this shouldn't be necessary for most people.

Andrew Lewis is a computer scientist and a relentless tinkerer, whose love of science and technology is second only to his love of all things steampunk.

Extreme Zap-a-Mole

Learn how a single microcontroller does the work of 20 old-school chips.

>> In my previous column I showed you how to use old-school logic chips to build Zap-a-Mole, an electronic version of the old Whac-A-Mole arcade game. Now I'm going to get rid of the old chips and substitute a microcontroller, which will make everything simpler while also enabling a lot of new features. This will give us Zap-a-Mole Enhanced. I'll then go even further to create a game so ambitious, it can only be known as Zap-a-Mole Extreme.

The PICAXE 28X1 is my choice of microcontroller. The "PIC" part of its name tells us it's a Programmable Interface Controller by Microchip Technology. The "AXE" was whimsically added by Revolution Education, which created a version of the BASIC computer language to control the chip. I like it because it's cheap and simple. (If you find that you enjoy playing with it, you may want to move up to the Arduino, which has many more features.)

Enhanced Zapping

Take a look at the program listing in Figure C. I deliberately squeezed its formatting, and there's a much more readable version online at makezine.com/24/electronics. Still, the tiny piece of code is all that the PICAXE actually needs to play the game. The question is, how do we get the instructions into the chip?

First, go shopping. From SparkFun Electronics you can obtain a PICAXE 28X1, a USB programming cable, and a mini audio jack. You also need an LCD display screen, such as the one sold with a driver by Peter H. Anderson (see Materials). Your total outlay will be around $50, but you can reprogram and reuse the microcontroller almost indefinitely. Check mouser.com or newark.com for other components.

Once you have your components, carefully follow the schematic in Figure B, either using separate buttons and LEDs, or buttons with LEDs inside them as suggested in my previous column. Figure A shows the circuit breadboarded, with the LCD screen substituting for the 7-segment numerals I used in

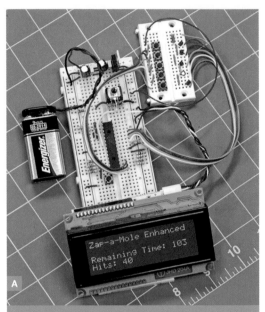

Fig. A: The breadboarded project is extremely simple compared with the version using logic chips pictured last issue.

the old-school version. The number of components and lengths of hookup wire have been drastically reduced.

Figure D (page 148) shows the pinouts of the 28X1. (Some pins have additional functions, which I've omitted for simplicity.) You'll see that the chip imposes its own pin-numbering system, so that in a program, pin3 means "input number 3," which is actually assigned to hardware pin 14.

Note that the Reset pin must be held high through a 4.7K resistor. Pins 6 and 7 connect with your computer through a mini stereo jack and a pair of resistors as in Figure E (page 148). Pins 8, 19, and 20 require a regulated 5-volt supply; a 9-volt battery can be used with an LM7805 regulator, and the

Photography and diagrams by Charles Platt

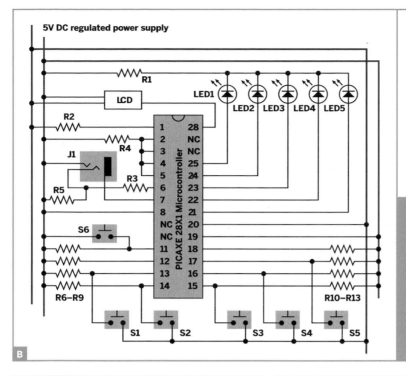

5V DC regulated power supply

COMPONENT VALUES
R1: 220Ω
R2: 4K7
R3: 22K
R4–R13: 10K
LED1–LED5: Generic
     5mm LEDs
S1–S6: SPST normally
     open push buttons
J1: 3.5mm mini
     stereo jack
LCD: Liquid crystal
     display (4-line,
     20-column,
     4,800-baud)

Fig. B: The Zap-a-Mole Enhanced schematic using a PICAXE 28X1 microcontroller.
Fig. C: All the code the microcontroller needs to play the game. (A fully commented, properly formatted version is available online at makezine. com/24/electronics.)

B

MATERIALS

For Zap-a-Mole Enhanced:
PICAXE 28X1 microcontroller IC SparkFun
     Electronics part #COM-08352, sparkfun.com, $10
USB programming cable SparkFun #PGM-08312
3.5mm stereo audio jack SparkFun #PRT-08032
PICAXE Program Editor (Windows) or AXEpad
     (Mac/Linux) software free from rev-ed.co.uk/
     picaxe/software.htm
Breadboard and hookup wire
LCD display screen, 4-line, 20-column, sold with
     an LCD117 driver at phanderson.com/lcd106/
     lcd107.html. Scroll down to 4,800-baud version.
Resistors: 220Ω (1), 4.7kΩ (1), 22kΩ (1), and 10kΩ
     (10) All can be 5%, and can be ¼-watt or ⅛-watt.
Illuminated push buttons (5) such as E-Switch
     part #LP4OA1PBBTR (that's oh-A-1, not zero-A-1)
Push button, SPST, normally open (1) Or use 6 of
     these if you're not using the 5 illuminated buttons.
LEDs, 5mm (5) if you're not using illuminated buttons
9V battery and battery snap
5V power regulator IC, LM7805 type
Capacitors, minimum 12V: 100µF electrolytic (2),
     and 0.1µF ceramic (2)
Wire strippers
Optional: Needlenose pliers, magnifier, multimeter

Additional components for Zap-a-Mole Extreme:
Decoder IC, CD74HC4514 type
Resistors: 150Ω, 1% (16); and 100kΩ (1)
     The 150-ohm resistors must be 1% tolerance.
Illuminated push buttons (11) Or use 11 more SPST
     push buttons and 11 LEDs.

```
setfreq m8 : high 7 : settimer t1s_8
pause 12000 : gosub screen1
ready: ' ----------------------------------------
outpins = 128
do : random w8 : loop until pin0 = 1
gosub screen2
w9 = 0 : w6 = 25000 : b6 = 180 : b3 = 0 : timer = 0
chooseled: ' ---------------------------------------
random w8 : b2 = w8 / 13107
if b2 > 4 or b2 = b1 then chooseled
low b1 : high b2 : b1 = b2 : w7 = 0
if b6 > 0 then : w6 = w6 - b6 : b6 = b6 - 1 : endif
checktime: ' --------------------------------------
if b3 = timer then getbutton
b3 = b3 + 1 : b4 = 120 - b3 : gosub screen3
if b4 = 0 then ready
getbutton: ' ---------------------------------------
b5 = 5 : b0 = pins
if b0 = 4 then : b5 = 0 : endif
if b0 = 8 then : b5 = 1 : endif
if b0 = 16 then : b5 = 2 : endif
if b0 = 32 then : b5 = 3 : endif
if b0 = 64 then : b5 = 4 : endif
newscore: ' ----------------------------------------
if b5 <> b2 then delay
w9 = w9 + 1 : gosub screen4 : goto chooseled
delay: ' --------------------------------------------
w7 = w7 + 100 : if w7 < w6 then checktime
goto chooseled
screen1: ' ------------------------------------------
serout 7, T4800_8, ("?fZap-a-Mole Enhanced") : return
screen2: ' ------------------------------------------
serout 7, T4800_8, ("?y2?x00Remaining Time: 120")
serout 7, T4800_8, ("?y3?x00Hits: 0   ") : return
screen3: ' ------------------------------------------
serout 7, T4800_8, ("?y2?x16", #b4, " ") : return
screen4: ' ------------------------------------------
serout 7, T4800_8, ("?y3?x06", #w9) : return
```

C

PICAXE manufacturer recommends 4 capacitors as in Figure F. You can mount them at the top of a breadboard as in Figure G.

Now install the PICAXE Program Editor (Windows) or AXEpad (Mac/Linux). My book *Make: Electronics* offers detailed instructions for setting everything up, or you can check the datasheet (see Sources, page 150). Finally, go to makezine.com/24/electronics and view the Zap-a-Mole Enhanced program, then copy and paste it into the Program Editor. Now download it to your PICAXE chip.

The game should start running automatically — when an LED lights up, the player must push the corresponding button to score — but with some important new features. It allows only a limited time for the player to press a button, and that time gets shorter as the game continues. It won't let you cheat by pressing more than one button at a time, and the screen displays your remaining playing time as well as your score. The online program listing includes comments telling you how to change these features, just by retyping a few numbers.

Extreme Sixteen

To make the game "extreme," I wanted more LEDs and buttons. Sixteen seemed a good number, in a 4×4 grid. The 28X1 doesn't have enough inputs and outputs, but there are ways around this limitation, and the techniques I'll describe may be useful to you in other applications in the future.

For the outputs, we can add a 74HC4514 decoder chip, which has 16 output pins, each able to drive a single LED (Figure H). Its 4 input pins receive a binary-coded number. This means, for example, that if the input pins have states 0010 (where 0 means low and 1 means high), the chip activates the LED attached to output 2. Similarly, pin states 0101 will activate output 5. To understand this relationship, I encourage you to read an introduction to binary arithmetic, which is fundamental in computing (see Sources, page 150).

Conveniently, the PICAXE can create binary patterns in its output pins that are exactly compatible with the decoder's input pins. For instance, the statement outpins = 5 will assign value 0101 to the first 4 output pins. So, we can connect the PICAXE directly with the decoder chip.

Now, how do we receive inputs from 16 push buttons? One method is by using a resistor ladder. Figure I shows a ladder consisting of 5 resistors. If

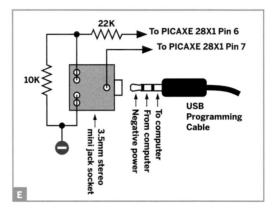

Reset	1		28	Output 7
ADC0	2		27	Output 6
ADC1	3		26	Output 5
ADC2	4		25	Output 4
ADC3	5		24	Output 3
From computer	6		23	Output 2
To computer	7	PICAXE 28X1	22	Output 1
Negative Power	8		21	Output 0
Resonator (optional)	9		20	Positive Power
Resonator (optional)	10		19	Negative Power
Input 0	11		18	Input 7
Input 1	12		17	Input 6
Input 2	13		16	Input 5
Input 3	14		15	Input 4

D

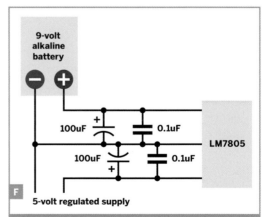

E

F

Fig. D: Pinouts of the PICAXE 28X1 microcontroller (some functions omitted for clarity). Fig. E: The PICAXE receives its program via a USB cable that terminates in a 3.5mm audio plug. Resistors must be included as shown, even when the plug is disconnected. Fig. F: Manufacturer's recommendation for a 5V regulated power supply. An AC/DC converter may be substituted for the battery.

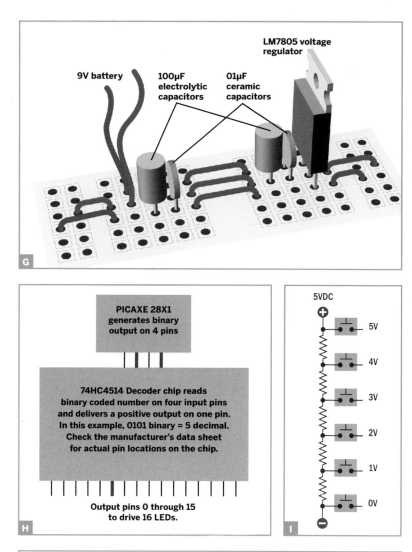

9V battery

100μF electrolytic capacitors

01μF ceramic capacitors

LM7805 voltage regulator

G

Fig. G: Installation of a 5V power supply at the top of a breadboard. Fig. H: Basic principles of the 74HC4514 decoder. Fig. I: With a resistor ladder, each push button supplies a different voltage. Fig. J: A ladder of 16 resistors enables one ADC pin to identify each of 16 push buttons. The resistors must have a 1% tolerance to create an even spread of voltages. All are 150Ω, except for the 100K pull-up resistor.

PICAXE 28X1 generates binary output on 4 pins

74HC4514 Decoder chip reads binary coded number on four input pins and delivers a positive output on one pin. In this example, 0101 binary = 5 decimal. Check the manufacturer's data sheet for actual pin locations on the chip.

Output pins 0 through 15 to drive 16 LEDs.

H

5VDC

5V

4V

3V

2V

1V

0V

I

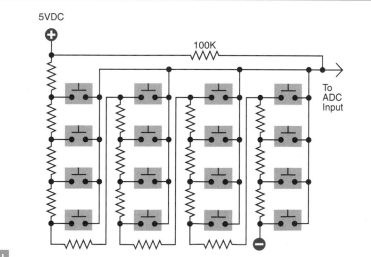

5VDC

100K

To ADC Input

J

More With Your Moles

Is that it? Not really. You could make a snazzy, playable enclosure for the game (Figure K.) And there are always more things to think about.

You could look up matrix encoding, for instance, to find an alternate way of sensing multiple devices with a smaller number of inputs, or driving multiple devices with a smaller number of outputs. Matrix encoding is an important topic, because it's used so often — for example, to address memory locations inside a computer.

You can also find matrix-encoded 4×4, 16-key keypads available cheap online. Maybe you could use one for the game? You could buy surface-mount LEDs and glue one to each button on the keypad. Getting voltage to them could be tricky — but I leave that to you.

How about adding a feature to adjust the game difficulty? The running speed could be controlled using a potentiometer hooked up to another ADC input.

And lastly, no game is complete without audio. The PICAXE can handle this, too; check the sound command in the BASIC manual (link below).

All of these improvements are enabled by the flexibility and power of a microcontroller. It can substitute for dozens of old-school logic chips, and if you take some time to learn the language that it understands, you can make it behave almost like a real computer. Not bad for a component that costs around $10.

Sources

» Program listings and schematics for Zap-a-Mole Enhanced and Extreme:
makezine.com/24/electronics
» Data sheet for CD74HC4514 (PDF):
makezine.com/go/decoder
» Binary arithmetic primer:
makezine.com/go/binary
» PICAXE installation and other info (PDF):
makezine.com/go/picaxe
» PICAXE BASIC manual (PDF):
makezine.com/go/picaxebasic

these all have the same value, and we apply 5V at the top and 0V at the bottom, we can tap into the ladder with push buttons and extract the voltages shown. If we have more resistors — 16, for instance — each step can be ⁵⁄₁₆ of a volt.

The 28X1 has some ADC inputs, each of which is connected internally with an analog-to-decimal converter. The chip senses voltage applied to a pin and converts it into a digital value from 0 through 255. If we make a ladder with 16 evenly spaced steps, using 150-ohm resistors, and we tap into them from 0 volts upward, the PICAXE will convert the voltages to values 0, 15, 31, 47, 63, and so on, up to 255.

Actually it's a little more complicated than this, because we can't allow the ADC input pin to "float" in an unconnected state when no button is pressed. I added a 100K resistor to take care of this (Figure J, previous page).

How can all of this be incorporated into the existing program? Really, you need a new program, and I already wrote it for you under the title Zap-a-Mole Extreme, available for download along with a schematic and parts list.

Charles Platt is the author of *Make: Electronics*, an introductory guide for all ages. A contributing editor to MAKE, he designs and builds medical equipment prototypes in Arizona.

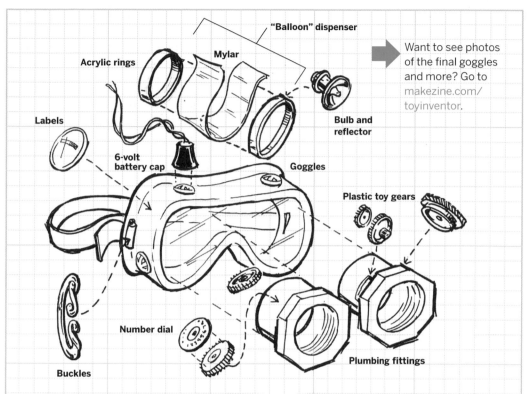

"Balloon" dispenser

Acrylic rings

Mylar

Want to see photos of the final goggles and more? Go to makezine.com/ toyinventor.

Labels

Bulb and reflector

6-volt battery cap

Goggles

Plastic toy gears

Number dial

Plumbing fittings

Buckles

When I received a last-minute invitation to a friend's bachelor party, I needed to come up with a gift idea, and fast. He was planning a steampunk-themed wedding, and I wanted to give him something to fit the occasion. You know the old design adage: speed, price, and quality — pick any two. Well, for once, this project came out fast, cheap, and great!

The gag gifts at the adult store were pretty lame, but the novelty glow-in-the-dark latex "balloons" had the right "wacky science" vibe. Aha! In a flash, I had my concept: Balloon-Dispensing Steampunk Honeymoon Night-Vision Goggles!

A pair of plastic safety goggles provided a sturdy and ready-made base on which to hot-glue various parts. A pair of eight-sided, threaded, plastic plumbing connectors along with some green-tinted cellophane became "night-vision" scope tubes. I glued on a handful of plastic gears, along with a number dial from a disposable camera, to add some interesting mechanical details. The rest I found in my broken-toy-parts bin.

To dispense the glow-in-the-dark balloons, I made an open-ended cylinder from a piece of clear mylar rolled and taped around two acrylic rings. A bulb and reflector scrounged from a penlight, with coiled wires and bogus connections, completed the night-vision look.

For a more Victorian feel I added some scrolled chrome buckles for the strap and Photoshopped a few labels to make dials and gauges as well as an appropriate nameplate.

To finish it off, I added touches of flat-black and metallic-brass spray paint — done, just in time for the party.

Bob Knetzger is an inventor/designer with 30 years' experience making all kinds of toys and other fun stuff.

Hawk Rescue!

The Scenario: You're one of a group of eight intrepid girl scouts, all 11 or 12 years old, who are off on their own for a hike in a mountain forest to prepare for a future overnighter. Suddenly you're all stopped by the sharp, plaintive peeping of baby birds. Searching for the source, you come upon a large tree beside a rocky mountain stream, at the base of which you find a mother hawk struggling with a broken wing, helpless to feed or defend her brood in the nest perched high above. It's clear that if you don't intervene, both the mother and chicks will soon fall prey to predators — and, given the location, hiking back out for help is not an option.

The Challenge: The nest is a good 25 feet off the ground, among a circle of sturdy branches, below which the tree is an almost perfect cylinder about 3 feet in diameter with no obvious way to climb. Still, all of you are determined to rescue this family and get them to the wildlife center near your base camp some five miles back down the mountain. Yet how can you possibly do that without endangering the birds or yourselves? Well, you are *scouts*, after all, no?

What You've Got: All of you are physically fit and range in height from 4 to 5 feet tall. You each have a backpack, hiking boots, a water bottle, and a 3-foot-square bandana. You also have among you a Leatherman tool, a Swiss Army knife, a basic first-aid kit, a roll of duct tape, and two four-man tents of ripstop nylon with the bungee cords and metal tent pegs that go with them. In addition, you have an emergency whistle and several pounds of trail mix. But all your cellphones were left back at the base camp (lest you spend the entire hike texting one another), and alas, *you have no cookies*. So now what?

Send a detailed description of your MakeShift solution with sketches and/or photos to makeshift@ makezine.com by Feb. 25, 2011. If duplicate solutions are submitted, the winner will be determined by the quality of the explanation and presentation. The most plausible and most creative solutions will each win a MAKE T-shirt and a *MAKE Pocket Ref*. Think positive and include your shirt size and contact information with your solution. Good luck! For readers' solutions to previous MakeShift challenges, visit makezine.com/makeshift.

Lee David Zlotoff is a writer/producer/director among whose numerous credits is creator of *MacGyver*. He is also president of Custom Image Concepts (customimageconcepts.com).

Photograph by Jen Siska

Photograph planets, record audio tracks, grab a tarp, build a shack, find true north, and learn for free.

TOOLBOX

Imaging Source USB, FireWire, and Ethernet Cameras

$350 and up astronomycameras.com

Whether you need a planetary camera for astronomy or a "grown up" webcam for machine vision experiments, you'll find Imaging Source's products appealing. These sub-megapixel USB, FireWire, and gigabit Ethernet cameras shine because of their versatility.

With a sturdy, metal-bodied construction, they're ideal for applications where a normal webcam would be too flimsy. The USB version requires no other power source. They work with all webcam software, and an especially versatile control program is included. All that's missing is audio — there's no microphone. And in some circumstances, that can be a blessing.

For astronomers, these are planetary cameras. The astronomy versions of the DMK (monochrome), DFK (color), and DBK (color plus infrared) cameras come with an adapter that fits a telescope eyepiece tube. Aim the telescope at a planet, record a few

minutes of uncompressed video, then use RegiStax (freeware) to sort and align the video frames, picking out the best so you get the full benefit of brief moments of steady air. These cameras will take still pictures and time exposures, but they're not designed for faint stars or nebulae.

For the rest of us, these are also general-purpose daytime cameras that accept C- and CS-mount video camera lenses. In this role they're great for machine vision, security camera applications, and general experimenting.

If you write your own software, you can use either the regular webcam interface or a more versatile set of software tools available from the manufacturer. You can control the exposure over a much wider range than with a webcam, and unlike a DSLR, there's no shutter to introduce vibration.

—Sharon Covington

» Want more? **Check out our searchable online database of tips and tools at** makezine.com/tnt**.**
Have a tool worth keeping in your toolbox? Let us know at toolbox@makezine.com**.**

RSVP Collapsible Silicone Funnel

$7 amzn.to/rsvpfunnel

Someone gave us this silicone funnel as a housewarming gift, and I looked cross-eyed at it before shoving it to the back of a drawer. But it turns out that we use it all the time. It's been used to funnel olive oil, homemade limoncello, beans, rice, and filling for a door sock. (It's also a big hit as a toddler toy.) The funnel is 4" in diameter and 4" tall when fully expanded, and it does in fact fit in the back of a drawer when collapsed. That gives it a big advantage over my old plastic funnel set, which practically needed a drawer to itself. It's also easy to clean. I'm thinking of buying a second one (in a fun color) to keep in the car.

—Arwen O'Reilly Griffith

Freescale Tower Microcontroller System

$99 freescale.com/tower

Got a project to prototype? The basic Freescale Tower kit consists of the 32-bit microcontroller board with USB and RS-232 interfaces, accelerometer, four display LEDs, switches, and a potentiometer. Additional Tower modules are available, including different microcontrollers, an LCD graphic display, sensor board, wi-fi board, and a prototyping board.

The kit's DVD contains the CodeWarrior IDE and supporting MQX software. It also includes 4 tutorial labs.

The Tower system is easy to set up and begin using. After experimenting with the labs, I was motivated to wire a solderless breadboard with light and temperature sensors to the Tower's side expansion port, then modify the lab code to read these sensors remotely via the internet. No soldering needed! Freescale's online documentation library makes project development easy.

—L. Abraham Smith, N3BAH

M-Audio Fast Track

$100 m-audio.com

I can remember a time when recording basic multitrack audio on a desktop computer was a daunting task and cost-prohibitive for most. Thankfully those days are behind us, and songwriters looking to record digitally have a number of affordable USB audio interfaces to choose from.

At about 100 bucks, the M-Audio Fast Track is one of the most affordable options in that category and (as it says on the box) aims to do one thing well: "record guitar and vocals on your computer."

The unit sports both ¼" instrument and XLR mic jacks (with 48V/phantom power) along with a ¼" stereo jack for headphones. Resulting audio quality and usability are quite nice, and many will appreciate the device's compatibility with the popular Pro Tools recording software (a limited version is included). Those looking to get acquainted with industry-standard DAWs will find this little number to be their "fastest track" to entry.

Oh, and if you have any trouble installing the included software on Mac OS 10.6.4 (like I did), be sure to check M-Audio's site for an update. —Collin Cunningham

Zeiss MiniQuick

$170 zeiss.com

If you've ever wanted to have a telescope with you at all times, you need the Zeiss 5x10 MiniQuick. I found it easier to hold steady than other monoculars because of its finger-friendly design, and the eyepiece to eye distance is comfortable for glasses wearers. The image is very sharp but not as bright as regular binoculars, making it unsuitable for night sky observing. Turned around, it also functions as a 2x–5x magnifier, although with some barrel distortion. It can also be an optical test instrument or component in a larger system. Since it's so lightweight, there's little excuse to leave the house without it. —SC

EZ Grabbit Premium Tarp Holders

$10 (4 to a package) grabbittool.com

I live in an old, leaky yurt. Given the rainy Northern California winters, this means I need a good tarp and a way to keep it tied down securely.

Thank goodness for these tarp holders. Each holder is made of two pieces, a "dog bone" and a "sleeve." The dog bone goes under the tarp; holding it stationary, you slide the sleeve over the tarp and the dog bone to secure it in place. The dog bone can be connected to a rope or mounted to a structure for more permanent placement.

You can use these holders anywhere on your tarp (not just on the edge), and you can move and reuse them easily. The 4"-long surface area of the holder creates a very secure hold on the tarp — much better than a grommet, which can rip out with too much tension.

EZ Grabbit Tarp Holders can also be used to connect two tarps together to make a larger one, or to connect the edges of a tarp to create a big bag for moving yard waste.

—Terrie Schweitzer

Fiskars 28-Inch Pro Splitting Axe

$31 amzn.to/splittingax

When you need an axe to split firewood, you want one that's sharp for good cutting and lightweight so it's easier on your arms and shoulders. This splitting axe from Fiskars fits the bill nicely.

The fiberglass composite handle keeps the axe light but makes it stronger than wood-handled axes. The 2¼lb axe head takes and holds an edge well.

I'm not a champion wood-splitter, but this ax makes it easy for me to split my own kindling, and my burly friends can use it to fly through a stack of wood in no time. —TS

Wood Blaster Log Splitter Wedge

$10 amzn.to/splitterwedge

If you want to perform amazing feats of wood splitting, then you need a splitting "grenade" and a sledgehammer.

This splitting wedge has a unique design that exerts force in multiple directions, so it works better than a more traditional splitting wedge.

To use it, just tap it into the end of a log and pound it in with the sledgehammer. You'll soon hear the satisfying crack of wood splitting.

This device makes a tough job much easier; you'll be amazed at the size of logs you can split. And because there are no sharp edges, using this tool feels a lot safer than using a splitting axe with a lot of force.

Best of all, when you show your friends how it works, they'll want to try it out themselves. You might feel a little like Tom Sawyer enticing his friends to whitewash the fence. —TS

BodyGuardz Protective Films

$15–$30 bodyguardz.com

I bought an iPhone 4. "Helicopter glass," I said in awe. But even helicopters get scratched up, so I never dared put anything else in my pocket but a hankie or a doggie doo bag. Nothing that could scratch my precious.

BodyGuardz gives you two sets of skins for your smartphone or iPad, plus lifetime replacements, for about 25 bucks. It's the same tough polyurethane film that 3M sells to protect car paint from rocks, so it's got a nice glassy shine and clarity. I can't tell any difference in touchscreen sensitivity.

Application is a bit fussy — you spray a soap solution to temporarily deactivate the adhesive — but unlike "dry-apply" films, it allows unlimited do-overs to get it placed just right on your precious.　　—*Keith Hammond*

Parallax RFID Read/Write Module

$50 bit.ly/RFIDreadwrite

Parallax has introduced a new RFID module, representing the first economical, fully assembled RFID reader/writer for hobbyist use. The device allows writing of up to 116 bytes of 32-bit password-protectable data using a simple 9,600-baud asynchronous serial connection.

The manufacturer's downloadable documentation provides full details on basic read/write commands. A "read legacy" command allows for reading of older style read-only tags. Command source code is downloadable for the BASIC Stamp 2 as well as Propeller object code. This makes it easy to serial interface with other microcontrollers such as Arduino.

The RFID read/write module is well constructed. The easy access and documented commands allowed me to integrate RFID into a hacked toy brainwave monitor I experimented with, so I could write the monitor's session data to a log on the RFID card via a BASIC Stamp 2 interface.
　　　　　　　　　　　　　　—LAS

Protoflex Adapter

$60 protoflex.net

Protoflex is a cool new electronic prototyping product. These IC component adapters are made from thin flexible PCB materials with an adhesive backing. All you do is peel and stick. Why didn't someone think of this before?

I assembled a prototype in only minutes by sticking the Protoflex adapters directly onto a thin plastic case without drilling holes or using a protoboard. It worked so fast I wanted to do it again.

Clever dual-hole solder patterns allow for throughhole or surface-mount wiring. The smaller hole pattern just fits a 30AWG wire and works great to hold it in place before soldering.

Protoflex panels come in SOIC, TSSOP/SSOP, SOT, SC-70, and DD-PAK IC devices plus adapters for Ribbon and D-Sub connectors. If you need to mount through-hole components, Protoflex even provides a special tool to align the flexible adapters to a 0.100" pitch protoboard. These adapters will probably change the way we all build circuits.

　　　　　　　　　　　　　　—*Tom Baycura*

« House Proud

Humble Homes by Derek Diedricksen
$19 relaxshacks.com

This 100% indie-produced book is a hoot, an education, and an inspiration, crammed with Derek "Deek" Diedricksen's cartoony designs — from practical to pie-in-the-sky — for "micro-houses," small backyard retreats, kid forts, treehouses, and other no-cost/low-cost outbuildings.

The emphasis here is on the fun and clever use of recycled materials. The book is obsessively illustrated with wonderfully wacky cartoons and design drawings or, as Deek describes it: "A carpal tunnel-inducing barrage of dime store pen sketches." It all brims with Yankee ingenuity, junkyard philosophy, and plenty of eye-rolling yucks.

The funky energy of *Humble Homes* reminds me a lot of the Lloyd Kahn *Shelter* books, Malcolm Wells' solar architecture books, and other handmade home books of the *Whole Earth Catalog* era. It's all hand-lettered and drawn, hand-assembled, GBC-bound, and the printing was "trash-funded" by selling recyclables. It doesn't get any more DIY than that!

—*Gareth Branwyn*

« Spin Doctor

Programming the Propeller with Spin by Harprit Singh Sandhu
$30 McGraw-Hill

This book is designed to transition users of traditional unicore microcontrollers such as Arduino and PICs to the opportunities and design concerns presented by the octo-core Parallax Propeller microcontroller and its high-level Spin programming language.

After explaining the hardware differences in parallel processors and their relationship to the Spin programming language, the author presents a series of basic experiments demonstrating pulse-width modulation, control of LCD and LED displays, and reading analog input.

This is neither a rehash of the official Propeller Manual, nor a substitute; in fact, it's a companion volume. Sandhu makes ample reference to Parallax documentation, providing deeper explanation of topics such as memory management, interfacing between the 3.3V Propeller and 5V peripherals, and the device's 32-bit counters.

While the projects are presented in the context of the Parallax Educational Kit, no special components beyond a Propeller chip are needed for basic experiments. Supporting devices such as a 2×16 LCD display, switches, resistors, and potentiometers would be found in the junk box of most readers. Advanced projects are presented as well, covering parallel processing in the use of stepper and servomotors and accelerometers.

This book should find a place on any Propellerhead's bookshelf, between Parallax's *Propeller Manual* and its *Programming and Customizing the Multicore Propeller* volumes.

—*LAS*

« Muscles Over Motors

The Human-Powered Home by Tamara Dean
$30 New Society Publishers

I've been playing around with the idea of making my own pedal-powered devices for a while now, and am waiting no longer now that I've come across this great primer. The book kicks off with an extensive and fascinating history of human-driven devices and "appropriate" technology, moves to an accessible data-driven section on the efficiency of various human-machine interactions, and gives a crash course on bicycle engineering for the layperson.

Throughout, Tamara Dean weaves in inspiring and ingenious uses of human power in countries all over the world, as well as current devices on the market covering everything from lawn mowers to ice cream makers and cellphones. In case you find yourself getting too wrapped up in 19th-century bicycle designs, the book culminates in detailed descriptions and illustrations of projects for adapting various devices to a standard pedal setup. Go get your salvaged exercise bike now! All in all, Dean makes a fantastic argument for putting your body where your needs are.
—*Meara O'Reilly*

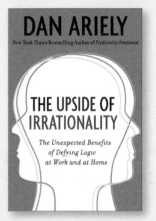

« A Thing Only Its Maker Could Love

The Upside to Irrationality by Dan Ariely
$28 Harper

"To increase your feelings of pride and ownership in your daily life, you should take a larger part in creating more of the things you use in your daily life," writes Dan Ariely in his entertaining and insightful new book.

My own DIY experiences (which I wrote about in my book, *Made by Hand: Searching for Meaning in a Throwaway World*) confirm what Ariely wrote, but Ariely actually conducted a series of experiments to arrive at this conclusion. He's a professor of psychology and a behavioral economist at Duke University, and this book (like his earlier bestseller, *Predictably Irrational*) explores the emotional side of human behavior regarding decisions about things of value.

The entire book is a joy to read, but the chapter about the satisfaction one gets from making stuff is especially interesting. Ariely's experiments on test subjects who were asked to make simple objects under a variety of conditions revealed "four principles of human endeavor": (1) Putting effort into making something not only changes the physical thing, it also changes the maker and the maker's evaluation of the object. (2) The more effort we put into making something, the more we love it. (3) Not only do we tend to overvalue things we make ourselves, we assume that other people will like them as much as we do. (4) If we can't finish a challenging project, we don't feel attached to it.

Ariely's conclusions now run through my mind as I make things, such as musical instruments. I really like my cigar box guitars (and I'm glad that I like them), but now I know that other people may not be as enamored of them as I am. That's OK! Other people should be making, and loving, their own stuff.
—*Mark Frauenfelder*

Edsyn FXF14 Fuminator Fume Extractor

$109 (not including filters) edsyn.com

A fume extractor is a simple device; it's just a fan that pulls air from your work area through a filter and out the back of the unit. Ever since Marc de Vinck made one that fits in a mint tin for around $10 (*see MAKE Volume 19, page 123*), the price tag on retail models has seemed unnecessarily high. That said, Edsyn's FXF14 (which is made in the USA) works great, has a small desktop footprint, and (in shiny, static-safe black) just plain looks good.

Edsyn claims its proprietary rotating filter system works eight times better than a stationary filter, and the pivoting head on the FXF14 makes it easy to get it close to your work. The fan is so quiet that sometimes I forget to turn the thing off.

I really enjoy mine and use it for soldering electronics as well as jewelry in my small apartment workshop. I usually feel a bit guilty after splurging on a fancy new tool, but your lungs can easily justify the investment of an effective (and attractive) benchtop solder-fume extractor.

—*Becky Stern*

Wikiversity Collaborative Learning

Free wikiversity.org

If I am nothing else, I am a student. For any skill I want to learn, I delve into books and websites, but unfortunately, not all skills can be learned without a classroom. Luckily, this problem is slowly disappearing with the introduction of the next level of Wikipedia — Wikiversity.

When I found this site by accident, I was elated. Normally, when I wanted to learn a new skill such as a foreign language, I was forced to find courses on local campuses that required tuition, or use online programs that cost several hundred dollars. With Wikiversity, I'm able to learn a language, practice with classmates, and use external references to keep learning.

The biggest drawback to Wikiversity is the lack of public awareness about it, which leads to a lack of participation. Some subjects aren't covered in great depth, or covered at all. The best way to rectify this is to spread the news about Wikiversity. I encourage everyone, whether you're a student or someone who has never been in a college setting, to give a Wikiversity course a try; you will be pleasantly enlightened.

—*Eric Ponvelle*

Iomega eGo BlackBelt Portable Hard Drive

$220 go.iomega.com

This excellent 1TB, 5,400rpm hard drive is not only the perfect portable drive, it's great for desktop storage. When on the road, you don't have to worry about dropping the device: Iomega claims the drive is shock resistant and able to withstand 84" falls onto industrial carpeting. That, plus its small profile and 9.6oz weight, make it a cinch to carry around.

My favorite feature is that it gets power from the data cable. Not having an AC wall wart reduces desktop clutter and further reinforces its convenience as a traveling drive. On the Mac edition there are two FW800 ports (FW800 to 400 cable included) as well as one USB 2.0 port, while the PC version features USB 3.0. Either way, daisy-chaining is a given. Bonus extras include downloadable security software as well as a 3-year warranty.

—*John Baichtal*

Tricks of the Trade By Tim Lillis

Track true north.

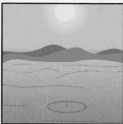

Trying to find true north but don't have a compass? Don't worry. As long as you know the time, you can find your way.

This trick works on a watch if you have one, but if you don't you can draw a clock face in the dirt or on some paper. Draw a circle, and within that, an hour hand pointing toward the sun.

Mark the current time where you drew your first line. Based on that, figure out where 12 would be and draw that in. If it's currently daylight savings time (summer) use 1 instead of 12.

Bisect the angle between the current time and 12, and draw another line. North will be on the far side of that line from the sun as long as you're in the non-tropical Northern Hemisphere.

Have a trick of the trade? Send it to tricks@makezine.com.

Titan X2 Flash Laptop Case

$310 amzn.to/x2flashlaptop or titanluggageusa.com

The X2 is a badass laptop case with a hard shell, a thick foam-padded interior, and enough extra organizational compartments inside that it could work as a briefcase. In fact, it's about the size of one and has a briefcase-style handle in addition to a removable shoulder strap.

Some people might get scared off by the hard shell, thinking it'll weigh a ton. In fact, the Flash weighs only 3.3lbs empty, so you'll be able to carry it easily on the bus. The hard shell offers great protection from ordinary dangers, but don't expect much more (don't let it get run over by a truck). Beyond the utilitarian angle, the Flash just looks sweet. Aside from a laptop, it's perfect for transporting nuclear codes, biological samples, or the blueprints for a giant laser — and even if you don't have anything like that to carry, the Flash totally looks the part. —JB

John Baichtal is a contributor to MAKE, makezine.com, and the GeekDad blog on wired.com.

Tom Baycura is an independent electrical engineering consultant with more than 15 years' experience in electrical and electromechanical prototype design.

Sharon Covington is aiming to be a science writer and the kind of person who knows exactly how to use both a pipette and a semicolon.

Tim Lillis is a freelance illustrator and avid DIYer.

Meara O'Reilly (mearaoreilly.com) is a sound designer, an artist-in-residence at the Exploratorium, and lives in a yurt-dome tent in Northern California.

Eric Ponvelle is a graduate of Nicholls State University in English with a concentration in technical writing.

Sean Michael Ragan is a jack-of-all-trades and master of, er — almost one.

Terrie Schweitzer lives in a yurt on a goat ranch in Sonoma County, Calif., and digs permaculture.

L. Abraham Smith, amateur radio licensee N3BAH, works with open-source hardware and software development every chance he gets, when not practicing law.

Have you used something worth keeping in your toolbox? Let us know at toolbox@makezine.com.

Sometimes it costs more to buy it than to make it from the money itself.

Photograph by Tom Parker

$17.95
Wood-frame Chinese abacus bought online.

↑ $0.93
Chinese abacus made of pennies, twigs, and coat-hanger wire.

The Chinese abacus, or counting frame, is an ancient calculator using beads on rods for both decimal and hexadecimal computation. In many parts of the world, schoolchildren still use these ingenious devices to do large-scale addition, sub-traction, multiplication, division, and even square- and cube-root operations at astounding speeds.

I made mine in the "Adirondack porch furniture" style, with a frame of dried chokecherry twigs. Since twigs aren't perfectly straight, I clamped each one in a jig so I could drill orderly and parallel holes using a drill press. First I drilled flat-bottomed holes for the joints of the frame, using Forstner bits. For the rods, I drilled carefully spaced sets of holes, sized so that I could press-fit coat-hanger wire. If you size your holes right, the frame and rods just snap together.

The traditional Chinese abacus has seven beads on each rod, two above and five below. These are usually made of wood, glass, or stone, rounded for easy manipulation. I wanted to make my beads out of pennies, but flat disks would be hard to separate and slide with your fingers. So I chose pre-1982 pennies, which are 95% copper and easy to deform. (The newer zinc pennies tend to snap when bent.)

I center-drilled each penny first. Then I fashioned a crude deforming tool using a bench vise. To one jaw of the vise, I glued an old cylindrical bushing the same diameter as a penny. I sawed off a 5⁄16" hex bolt, filed it to a smoothly rounded stub, and glued it to the other jaw, just opposite the bushing. This gave me a "tooth" I could use to deform each penny with a simple clamping operation.

MAKE's favorite puzzles. (When you're ready to check your answers, visit makezine.com/24/aha.)

That city is now the capital of an independent nation. Remove one letter from the country name and anagram the remainder to reveal the plural term for its currency.

Add one letter to the character name and anagram it to get the surname of an accused U.S. presidential assassin who lived for almost two years in the Soviet Union. In what "White Russian" city did he live?

The capital city belongs to an island nation where a famous hard-to-find children's character is called "Woh-ri." What is he called in America?

A neighboring country uses a currency of the same name. What's the surname of its current prime minister?

The style name contains the surname of a boxer who fought a bizarre 1976 match with wrestler Kanji Inoki that amounted to little more than Inoki kicking him in the legs for an hour. In what city was the fight?

That poet died in a Mediterranean country which gives its name to a style of calligraphy and a kind of type based on the calligraphy. What's the style called?

Add two letters to the prime minister's surname and anagram it to find one of that country's space programs. The two extra letters can be rearranged into the postal abbreviation of which U.S. state?

Add two letters to the state's name to get a different U.S. state, the birthplace of a U.S. president. In what town was he born?

The city's colonial name begins the two-word name of a dish of minced beef served in brown sauce. Anagram the second word to find the surname of an English poet. What kind of poem did he write about Melancholy?

Remove one letter from the town name and anagram it to make the name of a noodle soup. On what continent did it originate?

A southern African country played that game at its highest ("test") level from 1992 until political upheavals ended their capacity to field a competitive side. What is its capital?

The most popular sport in that country uses the same word to refer to the center of the field, a partnership, getting a man out, or a set of stumps. What word?

The continent name was used by the Romans to refer to one of their provinces, located on territory now belonging to two modern nations. Only one has ever hosted an Olympics. What is its capital?

A small southern U.S. city of the same name is home to a popular band whose first single in 1981 was named for an international broadcasting system founded during the Cold War. What's it called?

In January 2010, despite its name, that network began Pashto-language broadcasts in what densely populated country?

Circular Reasoning: Planet Earth

You're an astronaut from the Andromeda Galaxy, lounging about the Intergalactic Space Station. Another astronaut from the Boötes Dwarf Galaxy has just returned from Earth. Instead of lecturing about the geography, music, sports, politics, literature, history, and food of Earth, she decides to debrief her colleagues by presenting them with a puzzle about the small Milky Way Galaxy planet.

Each question refers to the answer to the previous question. But this puzzle's solution is a final large anagram for the first and last letters of all 15 responses, which reveals the astronaut's most important discovery.

Take the first and last letters of all 15 answers and anagram them here to fill in the 30 blanks, spelling a six-word sentence describing the expertise of this magazine's readers:

⬜⬜⬜⬜⬜⬜ ⬜⬜⬜⬜ ⬜⬜⬜ ⬜⬜ ⬜⬜⬜⬜⬜⬜ ⬜⬜⬜⬜⬜⬜⬜⬜⬜.

Carve a Stone Bowl

Make a gift to eternity in about two hours.

» Ever thought about trying to make something to last? It's a hard problem: moths, rust, fire, thieves, and time versus the works of the mighty.

Take a walk through an old cemetery. The Mount Auburn Cemetery in Cambridge, Mass., is one of my favorites. Acid rain is dissolving some of the marble and limestone headstones so that you can't even read them.

By contrast, bronze ornaments and plaques hundreds of years old look brand new. Unfortunately some of them are missing; they've been pried off by scavengers to sell for scrap. That's not a new problem — perhaps some of the bronze in those items came from da Vinci sculptures melted for cannon in Renaissance wars — hence old stone sculptures are more abundant than metal ones.

But even stone isn't safe. The looted pyramids, stripped of their limestone sheathing. The Sphinx's nose. The Venus de Milo's arms. Unfortunately, stone breaks rather than bending when accident or malice befalls it.

Marble statues in Greece were even crushed and fired down in limekilns to make mortar. Aztec temples in Mexico were broken up to build cobblestone streets and walls of churches. In my own hometown, a beautiful domed library was torn down to make room for a motel, the granite slabs and stones shipped off to build walls elsewhere. Only the granite columns remain, moved to a park.

So what lasts without being buried? Wherever I go in the world I see really ancient stone bowls and other funny little sculptures made from hard stone. They aren't big or square enough to split down and build a wall out of. They're somewhat useful and people keep them around. They can be made much thicker than ceramic items and are less likely to break.

For example, here's the worst-maintained museum in the world, in Fujian Province, China (Figure A). It's basically a vacant lot with a lot of artifacts piled in it. A collection of things that have withstood the

A

ravages of war, monsoons, and cultural revolution. Objects that won't rot, burn, melt, break, or be cut up to make something else of value. Notice the preponderance of stone bowls.

Now it's our turn. Let's also make something that will truly last. Fortunately, tools with diamond-studded cutters have become cheap and abundant. They make stone carving amazingly fast and easy. The same techniques seen here can of course be used to make any sort of stone objects you desire.

My bowl is heavy and shallow because I plan to use it for a mortar to make nut butter. And I want it to last forever.

MATERIALS AND TOOLS

Angle grinder	**Chisels**
Diamond abrasive disk	**Clamp**
Marking pen and	**Metal wedge I used**
tin cans	**an aluminum wedge**
Dust mask	**from the scrap bin.**
Safety goggles	**Big hard rock**
Hammer	

First, a few notes regarding materials. You can buy an imported *angle grinder* for a few dollars almost anywhere. When I travel in remote areas of the world, if there's electricity, there's an angle grinder.

Photography by Tim Anderson

An angle grinder that costs $10 new is worth exactly that. You'll keep buying better ones until you get one that lasts.

⚠️ **WARNING: The angle grinder is an incredibly dangerous and useful tool. Wear safety goggles, or better yet a full face shield and leather protective gear. An image search for "angle grinder injury" will scare the hell out of you. You'll want a guard on yours.**

(Unfortunately mine is missing its guard because I found it that way in the trash. The former owner threw it out when the switch burned out. The replacement switch cost a few dollars online.)

The *diamond abrasive disk* is a "dry cut segmented" masonry blade. Imported ones are available online for a few dollars, or wherever cheap tools are sold. Mine was $7. Goes through granite like you wouldn't believe. Opens up whole new worlds of recreational stone carving. Don't accidentally open your veins with it, OK?

Carve the Bottom
Let's get started. Does the stone "ring true"? Hang it from a cord and tap it all over (Figure B). The ear reveals a few spots where cracks and flaws deaden the sound. Nothing serious.

Mark a cutting line for a flat bottom of the bowl. A tin can or plastic tub flexes to make an approximately planar line around the stone (Figure C).

Clamp the stone securely, then cut around the

line with your diamond wheel (Figure D). Don't cut yourself. I did, fortunately not badly.

Chisel off the end of the stone, then grind the end of the stone flat.

Carve the Lip
Mark a cutting line for the lip of the bowl. I used this high-precision height gauge (Figure E). Clamp the stone, then cut all around the line as deep as you can.

Split the end off the stone with a wedge (Figures F and G). You just made a lid for your bowl!

Hollow the Bowl
Mark the inner edge of the lip of your bowl. Cut slots into the bowl (Figure H), and then break out the standing stubs of waste with a chisel or wedge. Repeat to achieve your desired depth.

Grind the Inside Smooth
Grind the inside of the bowl smooth with the edge of the grinding wheel (Figure I). You've made a bowl!

This article shows the first bowl I ever made. It took two hours according to the time stamps on my photos, and most of that was taking pictures and various distractions. Who knew stone carving could be so easy! Diamonds really are a girl's best friend after all, and the rest of us, too.

Enjoy your bowl, eternity! Hmm. How'd that be for the inscription?

Tim Anderson (mit.edu/robot) is the co-founder of Z Corp. See a hundred more of his projects at instructables.com.

Alessandro Volta and Electrodeposition

» Before Alessandro Volta invented the battery, the only way to produce electricity on demand was to use a friction machine and a Leyden jar — and except for parlor tricks, there wasn't much use for this form of electricity.

The problem was that the electricity generated in this fashion produced a single, big, instantaneous spark that, while entertaining (see "Remaking History" in MAKE Volume 21), wasn't usable in any practical way.

In 1799, Professor Volta was hard at work in his laboratory at the University of Pavia in Italy, investigating his idea that under certain conditions, dissimilar metals placed next to each other could generate an electrical current.

If this was so, he reckoned, then scientists could produce constant, even flows of electricity. Volta stacked alternating metal plates of zinc and copper. Between each group of plates he inserted disks of cardboard moistened with a weak saline solution.

This assembly became known as the *voltaic pile* — the first device that produced a steady source of electrical current. (An alternate configuration was his *couronne de tasses*, a "crown of cups" of saline solution linked by copper and zinc strips.)

When Volta published his findings in March 1800, the excitement in the scientific community was palpable. Sir Humphry Davy called the voltaic battery "an alarm-bell to experimenters in every part of Europe."

Soon, scientists were applying constant-current electricity to just about everything they could, eagerly studying and documenting the effects. Ultimately the scientific unit of electromotive force, the volt, would be named for Volta.

Less than six weeks after Volta wrote to the president of London's Royal Society with his findings, two British chemists, William Nicholson and Anthony Carlisle, were able to decompose water into its two component gases by electrolysis, applying electricity from a battery based on Volta's design.

> Before the chemists of the 19th century understood electrochemistry, most metal items were made from one solid hunk of metal; the same on the inside, outside, and everywhere in between.

This discovery was soon followed by Davy's production of electric light from a continuous spark in the gap between electrodes (see "Remaking History," MAKE Volume 20), and by perhaps the most commercially important of all early electrochemical applications: the electrodeposition of metals.

Before the chemists of the 19th century understood electrochemistry, most metal items were made from one solid hunk of metal; the same on the inside, outside, and everywhere in between. Most silver forks and knives, for example, were solid sterling silver, making them very expensive.

Using Volta's battery, with small amounts of metal salts and the right catalysts, it was possible to electroplate inexpensive metals with a thin, hard, and economical coating of pure nickel, copper, silver, or gold.

The Italian chemist Luigi Brugnatelli, a friend and colleague of Volta, invented electroplating in 1805, successfully plating silver medals in gold. The English caught on a few decades later, and by mid-century factories were mass-producing immense quantities of silver-plated teapots, silverware, hairbrushes, snuff boxes, and more, making life a little more refined, even luxurious, for the working classes.

In this issue's project, we'll use Volta's invention to do our own copper electroplating, using materials found in most hardware stores.

Make a Copper-Plated Sign

1. Cut 1×3 strips of copper and brass, then drill a
³⁄₁₆" hole in each as shown in Figures A and B.

2. Prepare the wood block by drilling holes as shown
in Figures A and C, so that the electrodes will be
suspended on opposite sides of the loaf pan. Next,
cut and bend two 12" copper wires as in Figure A.

3. Measure 6oz (170g) of copper sulfate (Figure D)
and mix it into 24oz (700ml) of water in a clean
bucket, bowl, or large measuring cup. Stir until
dissolved, but don't worry if a small amount
doesn't dissolve.

⚠ CAUTION: **Put on your rubber gloves and
splash-proof eye protection for the next 2 steps
and all steps thereafter that require you to
handle the acid solutions.**

4. Slowly add 3oz (89ml) of concentrated sulfuric
acid to the copper sulfate solution (Figure E). Be

Photography by Ed Troxell

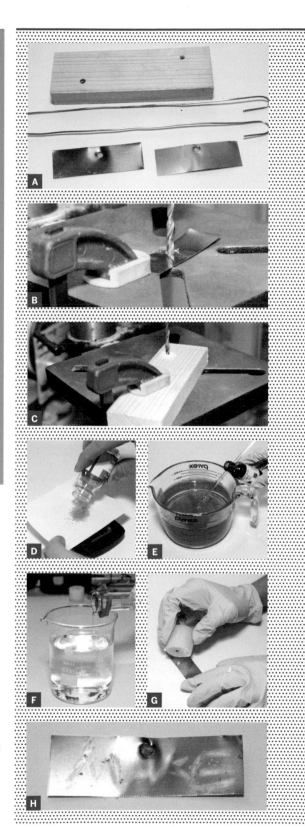

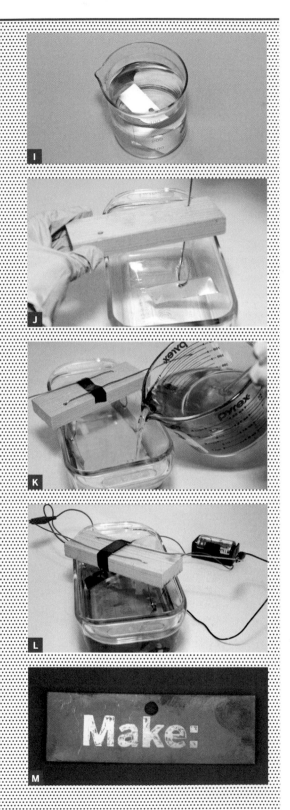

careful with this stuff — it is corrosive and highly toxic. Set aside.

5. Pour 350ml of water into the beaker. Slowly and carefully add 35ml of sulfuric acid to the beaker (Figure F, previous page). Stir with the glass stirring rod.

6. Thoroughly clean the brass and copper strips with dishwashing detergent and water. Rinse well and dry thoroughly. Use the wax candle to draw a design on the brass strip (Figures G and H). You can also try stenciling a design in wax.

7. Immerse the brass strip in the beaker of diluted sulfuric acid (Figure I). Remove it after 2 minutes. The acid prepares the metal for plating by removing oxides and other impurities. Electroplating professionals refer to this operation as "pickling."

8. Insert the wire hooks into the holes in the wood block and bend the wire so it lays flat on the block, taping the wire to the block if necessary. Then hang the brass and copper strips from the hooks (Figure J). The side of the brass strip with the wax design must face the copper strip.

9. Pour the copper sulfate solution into the glass baking pan. Don't overfill. Then place the wood block across the glass pan. Make certain the copper and brass strips are fully immersed in the copper sulfate solution (Figure K).

10. Use the alligator clip leads to connect the copper plate to the positive battery terminal and the brass plate to the negative battery terminal. You're electroplating! Let the electricity flow for approximately 2 minutes (Figure L).

11. Remove and rinse with lots of water, then rub with a soft cloth to polish off any wax traces. Your electroplated sign is complete (Figure M).

12. To dispose of the solutions, don't pour them directly down a storm sewer. Run cold water in your sink or bathtub and pour the solutions in, flushing them down with plenty of running water. Clean all the glassware with running water.

William Gurstelle is a contributing editor of MAKE.

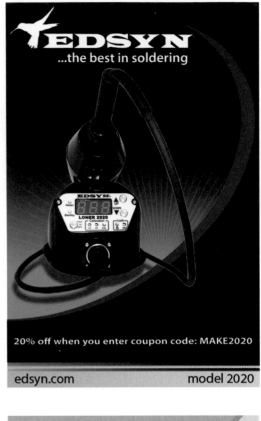

My first brewing experience came via a
Mr. Beer kit. The result, something akin to brownish vinegar, left much to be desired. I knew I needed better equipment, so I started with a few plastic buckets and a turkey fryer. My results improved, and I was hooked, but not satisfied. I wanted a system that allowed me to make beer from scratch — no more extracts, no more Mr. Beer.

I spent a few weeks formulating ideas and sketching out the mad science it would take to give life to my newfound passion. I developed a parts list as I went along, and scoured thrift stores and badgered friends for the necessary components: essentially, two pots or kettles to brew the beer in; a frame or stand to hold the kettles; a way to heat the beer; and a pump to circulate the contents through the system. An old steel table became the base for the stand, and abandoned kegs from a local liquor store served as my kettles. The table formerly had a glass top, so I had to fill in the gaps with scrap metal and use more scraps to complete the basic framework. With some amateur welding and a fresh coat of paint, I soon had the infrastructure in place.

Next, I needed some heat. With my old system, I'd used propane to boil the wort (unfermented beer). I wanted my "FrankenBrewery" to run on electricity,

not only for consistency in temperature but also in case I ever found a way to power it with the sun. My solution divided the system into two parts: high-voltage and low. The front control panel was designed to run on 12V DC, and a sealed NEMA box housed the 240V AC power for the heating elements. For temperature control, I adapted a PWM circuit to vary the power to the elements.

Next, I developed a way to move my brew from kettle to kettle. The use of an electric ball valve allowed me to remotely control flow from the pump. Using a mix of copper and PEX tubing for the main routes, I incorporated a quick-change panel on the front to help move fluid around using only one pump. I was ready for a test run.

On a frigid Colorado afternoon, I hooked up the power and started flipping switches. Lo and behold, rather than the lights flickering and the machine rising up and smashing through the door, it worked like a charm. Since the inaugural run, several delicious batches of beer have flowed through my creation, with the process becoming more familiar and more fulfilling each time.

Matthew Wirtz of Eagle, Colo., is soon to be co-owner of Bonfire Brewing Company.

Photograph by Matthew Wirtz